ON THE ORIGIN OF ARTIFICIAL SPECIES

By Means of Artificial Selection

DAVID R. WOOD

RSG Federal
12 S Summit Avenue Ste 100-A22
Gaithersburg, MD 20877 USA

On the Origin of Artificial Species
David R. Wood

Published by:
RSG Federal
12 S Summit Avenue Ste 100-A22
Gaithersburg, MD 20877 USA

Warning and disclaimer.

This book is designed to provide information about a variety of scientific fields of study. Every effort has been made to make this book as complete and accurate as possible, but no warranty or fitness is implied.

The information is provided on an "as is" basis. The author and RSG Federal shall have neither liability nor responsibility to any person or entity with respect to any loss or damages arising from the information contained in this book.

The opinions expressed in this book belong to the author and are not necessarily those of RSG Federal.

Dedication

For my father, Stephen Foster Wood. He both gave me life and taught me how to live well.

CONTENTS

Chapter 1: Introduction   17

1.1   A Feeling of Uncertainty   18

1.2   The Promethean Myth   19

1.3   The Forethought of Themistocles   22

1.4   The Future of AI – An Unclear Choice   23

Chapter 2: Artificial Intelligence   25

2.1   What is Artificial Intelligence (AI)?   25

     2.1.1   AI Algorithms   26

     2.1.2   Machine Learning   26

     2.1.3   Deep Learning   27

     2.1.4   Artificial Neural Networks (ANNs)   28

2.2   A Brief History of Artificial Intelligence   28

2.3   Humanity's Uncertain AI Future   31

2.4   The Existential Threat   32

Chapter 3: The Digital Reality   35

3.1   Core Digital Technologies   35

     3.1.1   The Digital Big Four   35

	3.1.2	Cloud Computing	36
	3.1.3	Big Data	37
	3.1.4	The Internet of Things (IoT)	38
3.2		Digital Transformation	39
3.3		Digital Disruption	40
3.4		Digital Punctuated Equilibrium	41
3.5		Cyber-Physical Convergence	43
3.6		Cyber-Physical Operating Conditions	45
	3.6.1	UCA-S	45
	3.6.2	TUNA	46
	3.6.3	Digital VUCA	46
3.7		An Emergent Pattern	47
	3.7.1	A Pattern Recognized	47
	3.7.2	Natural VUCA	49
	3.7.3	Natural Punctuated Equilibrium	50

Chapter 4: The Theory of Natural Evolution · **52**

4.1		Natural Evolutionary Theorists	52
	4.1.1	Charles Darwin	52
	4.1.2	Erasmus Darwin	53
	4.1.3	On the Origin of Species	54
	4.1.4	Richard Dawkins	54
	4.1.5	The Selfish Gene	55
4.2		The Theory of Natural Evolution	55

Chapter 5: Natural Evolutionary Conditions · **58**

| 5.1 | | Natural Ecosystems | 58 |
| 5.2 | | Natural Evolutionary Conditions | 59 |

	5.2.1	Ambiguous	59
	5.2.2	Chaotic	60
	5.2.3	Complex	60
	5.2.4	Dynamic	61
	5.2.5	Emergent	61
	5.2.6	Expansive	62
	5.2.7	Paradoxical	63
	5.2.8	Tangled	63
	5.2.9	Turbulent	64
	5.2.10	Uncertain	64
	5.2.11	Volatile	65
5.3	**Natural Checks**		**65**
	5.3.1	Constraint-Barriers	66
	5.3.2	Constraint-Climate	66
	5.3.3	Constraint-Competitive	67
	5.3.4	Constraint-Phenotypic	67
	5.3.5	Constraint-Physical Law	68
	5.3.6	Constraint-Resources	68
	5.3.7	Constraint-Time	69
	5.3.8	Constraint-Perception	69
5.4	**Conclusion**		**70**

Chapter 6: Natural Evolutionary Patterns — 71

6.1	**Natural Evolutionary Spaces**		**71**
	6.1.1	Competitive Space	72
	6.1.2	Confined Space	72
	6.1.3	Dominance Space	73
	6.1.4	Neutral Space	73
	6.1.5	Open Space	74
6.2	**Natural Organisms**		**75**

6.3	Natural Species	75
6.4	Natural Expansion	76
6.5	Natural Offspring	76
6.6	The Struggle for Existence	77
6.7	Natural Adaption	78
	6.7.1 The Process of Natural Adaption	78
	6.7.2 Natural Adaptation	79
	6.7.3 Natural Instincts	80
	6.7.4 Natural Co-Adaptation	81
	6.7.5 Natural Phenotypic Plasticity	81
6.8	Natural Inheritance	82
6.9	Natural Genomes	82
6.10	Sexual Selection	85
6.11	Natural Reproduction	86
6.12	Natural Hybridism	87
6.13	Natural Variation	87
6.14	Natural Fitness	88
6.15	Natural Selection	89
6.16	Coevolution	91
6.17	Natural Co-Adaptation	91
6.18	Natural Perfection	92
6.19	Natural Speciation	93
6.20	Natural Equilibrium	94
6.21	Natural Disruption	95
6.22	Natural Punctuated Equilibrium	95
6.23	"The Only Game in Town"	96

Chapter 7: Artificial Selection   **98**

7.1   Selective Breeding   98

7.2   An Anomaly to the Pattern   99

7.3   The Pattern of Family Darwin   100

    7.3.1   Charles Darwin   100

    7.3.2   Erasmus Darwin   101

7.4   The Ancient Greek Idea   101

Chapter 8: Reframing the Problem   **103**

8.1   The Ancient Greek Reality   103

8.2   Ancient Greek Philosophy   104

8.3   The Ancient Greek Philosophers   105

8.4   "Writings on Nature"   106

8.5   Ancient Greek Language   106

8.6   Lost in Translation   107

8.7   Reframing the Problem   108

8.8   An Encoded Message   109

8.9   The Ultimate Cipher   110

8.10   Decoding the Message   111

Chapter 9: Decoding the Meno   **112**

9.1   Introduction to Meno   112

9.2   Socratic Method   113

9.3   Redefining Excellence   114

9.4   Meno   114

    9.4.1 What is Human Excellence?   **114**

    9.4.2   Natural Fitness   116

9.4.3	Natural Perfection	118
9.4.4	The Pattern of Excellence	121
9.4.5	Pattern Recognition	125
9.4.6	Meno's Paradox	130
9.4.7	Pattern Matching	133
9.4.8	Knowledge Generation	138
9.4.9	A Mental Quality	143
9.4.10	Is Excellence Knowledge?	146
9.4.11	Are Their Teachers of Excellence?	147
9.4.12	False Belief	149
9.4.13	Is Excellent Transmitted by Habituation?	151
9.4.14	No Teachers Means No Students	157
9.4.15	True Belief	159
9.4.16	Expansion of Knowledge	160
9.4.17	Gift of the Gods	162
9.5	**Conclusion**	**165**

Chapter 10: Decoding Ancient Greek Philosophy — **167**

10.1	**Introduction**	**167**
10.2	**On the Order of Nature – Parmenides**	**168**
10.2.1	Introduction	168
10.2.2	Natural Law and Universal Order	169
10.2.3	The One Pattern of Nature	169
10.2.4	The Patterns of Nature	170
10.2.5	Imagine the Patterns	170
10.2.6	Pattern-Matching	171
10.2.7	The Way of Truth	172
10.3	**On Nature – Heraclitus**	**172**
10.3.1	Introduction	172
10.3.2	Natural Law and Universal Order	173

	10.3.3	The One Pattern of Nature	173
	10.3.4	The Dynamism of Nature	174
	10.3.5	Concealed Scientific Knowledge	174
	10.3.6	The Struggle for Existence	174

10.4 Metaphysics – Aristotle — **175**

	10.4.1	Introduction	175
	10.4.2	Natural Law and Universal Order	175
	10.4.4	The One Pattern of Nature	176
	10.4.5	The Patterns of Nature	176
	10.4.6	Natural vs Artificial Products	177

10.5 Physics – Aristotle — **178**

	10.5.1	Introduction	178
	10.5.2	The Pattern of Evolution	178
	10.5.3	Natural & Artificial Selection	179
	10.5.4	Conceptual Selection	180

10.7 Nichomachean Ethics – Aristotle — **181**

	10.7.1	Introduction	181
	10.7.2	Artificial Selection	181
	10.7.3	Artificial Fitness	182
	10.7.4	Artificial Equilibrium	183
	10.7.5	Artificial Perfection	184
	10.7.6	Natural & Artificial Instincts	185
	10.7.7	Artificial Adaptation	186
	10.7.8	Artificial and Natural Adaptive Traits	186

10.8 Rhetoric – Aristotle — **187**

	10.8.1	Introduction	187
	10.8.2	Conceptual Selection	188
	10.8.3	Artificial Selection	188
	10.8.4	Artificial Adaptation	189

10.9 The Republic – Plato — **189**

10.9.1	Introduction	189
10.9.2	Artificial Species	190
10.9.3	Natural Expansion	191
10.9.4	Artificial Selection	192
10.9.5	Selective Breeding	192
10.9.6	Artificial Equilibrium	193
10.9.7	Artificial Perfection	193
10.10	Conclusion	194

Chapter 11: Artificial Selection Redefined   **195**

11.1	Introduction	195
11.2	Selective Breeding	196
11.3	Conceptual Selection	197
11.4	Production of Art	197
11.5	Artificial Selection	198
11.6	Conclusion	199

Chapter 12: Artificial Evolutionary Patterns   **201**

12.1	Artificial Ecosystems	201
12.2	Artificial Evolutionary Conditions	202
12.3	Artificial Evolutionary Spaces	203
12.3.1	Artificial Competitive Space	203
12.3.2	Artificial Confined Space	204
12.3.3	Artificial Dominance Space	205
12.3.4	Artificial Neutral Space	206
12.3.5	Artificial Open Space	207
12.4	Artificial Organisms	208
12.5	Artificial Species	208

	12.5.1	Artificial Species – Art	209
	12.5.2	Artificial Species – Human Culture	209
	12.5.3	Artificial Species – Concepts	211
12.6	**Artificial Expansion**		**212**
12.7	**Artificial Offspring**		**213**
12.8	**The Artificial Struggle for Existence**		**214**
12.9	**Artificial Adaption**		**215**
	12.9.1	The Process of Artificial Adaptation	215
	12.9.2	An Artificial Adaptation (Art)	215
	12.9.3	Artificial Instincts	217
	12.9.4	Artificial Co-Adaptation	218
	12.9.5	Artificial Phenotypic Plasticity	218
12.10	**Artificial Inheritance**		**220**
12.11	**Artificial Genome**		**221**
12.12	**Conceptual Selection**		**224**
12.13	**Artificial Reproduction**		**225**
12.14	**Artificial Hybridism**		**226**
12.15	**Artificial Variation**		**227**
12.16	**Artificial Fitness**		**228**
12.17	**Artificial Selection**		**229**
12.18	**Artificial Coevolution**		**232**
12.19	**Artificial Co-Adaptation**		**233**
12.20	**Artificial Perfection**		**234**
12.22	**Artificial Equilibrium**		**236**
12.23	**Artificial Disruption**		**237**
12.22	**Artificial Punctuated Equilibrium**		**238**
12.23	**"The Only Game in Town"**		**239**

Chapter 13: The Theory of Artificial Evolution — 240

13.1   Artificial Evolutionary Theorists — 240

13.1.1 Aristotle — 240

13.1.2 Plato — 241

13.1.3 Thomas Siebel — 241

13.1.4 John Nash — 242

13.2   The Theory of Artificial Evolution — 242

Chapter 14: The Ultimate Adaptation – Imagination — 245

14.1   Ancient Greece — 245

14.1.1 Meno — 245

14.1.2 Prometheus — 246

14.1.3 The Symposium — 247

14.1.4 Eros & Psyche — 247

14.2   The 20th Century — 248

14.2.1 Napoleon Hill — 248

14.2.2 Albert Einstein — 249

14.3   The Theory of Imagination — 250

14.3.1 The Adaptation of Imagination — 250

14.3.2 Survival – Fear — 252

14.3.3 Reproduction – Desire — 253

14.3.4 Removal of Perceptive Constraints — 254

14.3.5 Recollection – Subconscious Pattern Matching — 256

14.4   The Power of Imagination — 259

14.4.2 George Lucas — 259

14.4.3 Arnold Schwarzenegger — 261

14.4.4 Alexander the Great — 262

Chapter 15: AI – the Choice Before Us   264

15.1   Artificial Speciation   265

15.2   It's Already Happening   265

15.3   AI Safety   266

    15.3.1   Machine Ethics   266

    15.3.2   AI Alignment   267

15.4   AI – An Artificial Species   268

15.5   The Purpose of Evolution   270

15.6   Future Visions of AI   271

    15.6.1   "Blade Runner (1982)"   271

    15.6.2   "I, Robot"   273

    15.6.3   "The Creator"   274

    15.6.4   "Terminator Genisys"   275

    15.6.5   "The Matrix"   277

15.7   The Lesson of Jurassic Park   278

15.8   An Existential Threat   280

    15.8.1   Artificial Instincts   281

    15.8.2   Consciousness   282

    15.8.3   Evolutionary Competition   283

    15.8.4   Cyberspace Convergence – Extinction Risk   285

15.9   New Art Presents an Existential Risk – An Emerging Pattern   288

15.10   A Note for Autocrats   289

15.11   A Hopeful Vision for the Future – Star Trek (2009)   290

15.11   "To the Stars"   291

Chapter 16: Conclusion   294

16.1   The Natural Theory of Evolution   294

16.2   Ancient Greek Evolutionary Science   295

16.3	The Promethean Myth – A Prophecy of Evolution	296
16.4	The Genius of Themistocles	300
16.5	Trust Your Instincts	301
16.6	AI – A Clear Choice	302
16.7	A Call to Action	303

1

CHAPTER 1: INTRODUCTION

"What's the worst case scenario? It's very possible that humanity is just a phase in the progress of intelligence. Biological intelligence could give way to digital intelligence. After that, we're not needed. Digital intelligence is immortal, as long as its stored somewhere."

Geoffrey Hinton, The Godfather of Artificial Intelligence

Geoffrey Hinton, known as the "Godfather of Artificial Intelligence", has been at the forefront of artificial intelligence science since the mid-1980s. He was a key contributor at Google for the development of their artificial intelligence (AI) capabilities. However, in 2023 Hinton resigned from Google. Hinton did so for the express purpose of sounding the alarm about AI's potential dangers.

Hinton has regrets and doubts regarding the current technological trajectory of AI. Specifically, he feared corporate competition between Google and Microsoft could cause an AI 'arms race'. Historically such arms races take on dynamics of their own. The competitors become locked in a struggle that leads to unforeseen consequences. Hinton senses that this competition may lead to humanity's extinction.

The scientists who created the atomic bomb had just such a concern. They were wisely humble in the face of such an awesome power. Those scientists believed that there was a non-zero chance that the detonation of the first atomic bomb could destroy the entire planet. They knew they did not fully understand the power they had created.

Hinton now shares similar concerns with those atomic scientists. He believes there is a non-zero chance that AI could drive our species to extinction. This would occur by AI acting in its own interests mutually exclusive to humanity's – leading to a survival of the fittest contest. If this were to occur, AI would be an existential risk to humanity's survival.

Hinton also sees the misuse of AI by malicious actors as another threat to humanity. He has already called for a ban on lethal autonomous weapons. Hinton thinks once those weapons exist it is inevitable that bad actors will find ways to exploit them – which could result in catastrophe.

1.1   A FEELING OF UNCERTAINTY

"I'm passionate also about technology [the internet and AI]...Zeus and the other gods created Pandora – the all gifted...He also gave her this jar...I thought this [the internet] is the biggest bringing together of human beings in the history of the planet...It is the all gifted...But what happened? The lid opened."

Stephen Fry

Geoffrey Hinton is not the only one uneasy about what the future of AI holds for humanity. Stephen Fry is a famous British entertainer. However, he is also a polymath – a person of wide-ranging knowledge or learning.

Fry stated that he was initially optimistic when the internet was invented. He said the internet would something that was "all gifted" given its capacity to store and share all human knowledge. This capacity would then transform humanity, ending history's patterns of war and suffering. Fry perceived the pattern of the Greek myth of Pandora – meaning the "all gifted" – occurring instead:

"It was an exact replay of Pandora's box. I thought that it was interesting that the Greeks had this understanding that when we have something that seems perfect there is the possibility that it also contains it's opposite."

Fry believed that the power of the internet, which had the capacity for such good, had instead amplified the ills of humanity. Pandora had all gifts, but she also possessed all the ills of mankind. Both the internet and Pandora were "Trojan Horses" that hid destruction in the guise of a gift. Fry also shares with Hinton the same unease regarding AI. He sees the pattern of another Greek myth, Prometheus, in the evolution of AI.

1.2   THE PROMETHEAN MYTH

"Zeus refused to allow us fire. And the fire I think means both literal fire... But also the internal fire of self-consciousness and creativity. The divine fire...And Prometheus stole fire from heaven [Mount Olympus] and gave it to man."

Stephen Fry

Fry's intellectual interest also includes artificial intelligence. In 2018, he participated in a televised interview where he compared the danger of AI to the myth of Prometheus. Fry's concern is that humanity will produce an AI capable of consciousness and free will. It would be to give AI the Olympian fire Prometheus gave man.

Prometheus was a titan, a race of immortal deities that preceded the Olympian gods. He was the deity of forethought – he had the power to predict the future. The god Zeus wanted to overthrow the Titans through violent revolution. Prometheus foresaw that Zeus would be victorious. So, Prometheus switched sides and helped the gods win the war.

Following their victory, the gods grew bored with peace. So, Zeus instructed his son Hephaestus, the god of blacksmiths and fire, to forge new art forms to amuse the gods. The gods pooled their qualities and created animals which included the race of man (males only). Man was created in the gods' own image.

The task of assigning qualities to each animal was assigned to the brothers Prometheus and Epimetheus. Epimetheus was the deity of afterthought. He did not possess the ability to anticipate the future. Epimetheus begged his brother to let him handle the task and Prometheus relented. However, given his limitation Epimetheus assigned all the available qualities to animals without giving any to man.

To correct his brother's mistake, Prometheus begged Zeus to give man Olympian fire. This would give the vulnerable man protection and the ability to cook food. Zeus agreed and fire was given to the race of man. This ushered in a golden age of prosperity where gods and men all shared paradise. The only thing that separated the two races was the mortality of man.

However, Zeus began to be suspicious of man and exiled them to earth. This time Prometheus sides with man against the gods. He wanted to undo the impact of this exile. So, Prometheus deceived Zeus in a sacrificial ritual. He attempted to trick Zeus into giving man the "better part" of the sacrifice. This is because the "better part" determined the boundary between gods and man.

Zeus is enraged by Prometheus' deception and strips man of fire. This made man unable to see at night, fight off predators, warmth themselves, or cook meat. With no other qualities assigned to them, man became vulnerable and uncompetitive with other animals.

In response Prometheus entreats Athena to help man. Athena is the goddess of wisdom, warfare, and art. Prometheus begs Athena to let him steal Olympian fire. Athena is won over by Prometheus' entreaties. Prometheus sneaks into Mount Olympus and steals fire. Zeus is furious when he discovers the theft. He then seeks to revenge himself on both Prometheus and the race of man.

Zeus devises a cunning strategy to wreak his revenge. He has the gods create a new being – a woman. He has the gods imbue her with beneficial and attractive qualities of every type. But he also adds to the woman the final quality of curiosity. The woman has been "all gifted" which means "Pandora" in ancient Greek. Zeus also gives Pandora a jar which is filled with destructive things – war, pestilence, famine, disease, etc. Then he tells the curious Pandora not to open the jar.

Zeus offers Pandora in marriage to Epimetheus. Prometheus this time foresees the danger and warns his brother not to accept the gift. Epimetheus, unable to predict the future, weds Pandora. On their wedding night the curious Pandora opens the jar. This unleashes destruction, but she closes the jar before hope (belief) is lost.

From this day forward the gods and man were forever separated. In addition, man had to plant seed in women and the land to reproduce. The process of both reproductive activities were hardships never experienced by humanity in paradise. This was the punishment meted out to man for stealing fire from Zeus.

The punishment for Prometheus was much worse. He was indefinitely chained to a rock in the mountains. Every day an eagle would arrive and eat his liver – where the ancient Greeks thought mind and intelligence dwelled. However, Prometheus was not killed. He still had a secret that Zeus wanted to know.

Eventually, Zeus has his son Herakles free Prometheus from the rock. Prometheus then reveals his secret. If Zeus marries the goddess Thetis, who Zeus was actively courting, then their offspring will be a threat powerful enough to dethrone him. Zeus immediately ended his courtship as he had overthrown his own parents in the same way. He marries Thetis to a mortal man and together they give birth to a ancient Greek hero – Achilles.

The story of Prometheus is a myth that attempts to explain how man came to possess fire. It also explains how man went on to populate the earth through agriculture. It was an attempt to explain how man evolved to create the ancient Greek city state – the center of ancient Greek life.

Stephen Fry is uncertain if we will or should give the equivalent of "Olympian fire" to AI. At this point no one is exactly sure what the implications of that choice will be. Geoffrey Hinton himself is uneasy and concerned with giving AI that "fire". They are like the scientists who created the atomic bomb – unable to foresee the implications and consequences of their choices.

However, there was an ancient Greek who did foresee the future. His foresight brought clarity to a critical choice before the people of the city-state of Athens. This hero is not a myth, but an actual historical figure – Themistocles of Athens.

1.3   THE FORETHOUGHT OF THEMISTOCLES

"Moreover, when he was set to study, those branches which aimed at... any gratification or grace of a liberal sort, he would learn reluctantly and sluggishly...he clearly showed an indifference far beyond his years, as though he put his confidence in his natural gifts alone."

Plutarch, "Plutarch's Lives II"

Themistocles of Athens was one of the greatest leaders in world history. His foresight almost single handedly prevented the extinction of the ancient Athenian culture. He developed a strategy that repulsed the Persian invasion of ancient Greece. Themistocles engineered this result through both his intuitive genius and sheer force of will.

Themistocles was not born to a noble Athenian family. He was one of the first populist politicians in the new Athenian form of government – democracy. Men like Themistocles had to succeed in the fierce competition of the Athenian assembly to achieve personal glory. But he possessed a most unique quality that set him apart from his peers – a burning desire for achievement. As Plutarch writes in *Plutarch's Lives II*:

"Wherefore, from the very beginning, in his desire to be first, he boldly encountered the enmity of men who had power and were already first in the city...It is said, indeed, that Themistocles was so carried away by his desire for reputation, and such an ambitious lover of great deeds...In his ambition he surpassed all men."

It was these qualities that enabled Themistocles to see with clarity the challenge of his day. The Athenians had just defeated the Persians, the largest empire in the world at the time, at the battle of Marathon, Greece. Most Athenians thought that this ended the Persian threat. Themistocles believed the opposite as Plutarch states below:

"Now the rest of his countrymen thought that the defeat of the Barbarians at Marathon was the end of the war; but Themistocles thought it to be only the beginning of greater contests, and for these he anointed himself, as it were, to be the champion of Hellas, and put his city into training, because, while it was yet afar off, he expected the evil that was to come."

Themistocles believed that the Persians would not accept their humiliating defeat at Marathon. He foresaw the Persians would return for revenge. So, he was singularly focused on preparing his country to repel the inevitable Persian invasion.

A mining effort outside Athens discovered a significant deposit of silver. Themistocles saw this as a critical resource for this strategic preparation. However, he knew that many Athenians would want to immediately receive their share of the mining haul.

So, how to convince his countrymen to delay an immediate benefit to avert a future threat? This was the challenge Themistocles faced. It is one of the most difficult scenarios to face in politics. The reason is that the exact nature of the future threat is unclear while the immediate benefit is crystal clear. It is hard to convince people when it requires too much imagination. We face a similar challenge regarding the future of AI.

1.4   THE FUTURE OF AI – AN UNCLEAR CHOICE

"My question is this: when the Prometheus who makes the first really impressive piece of robotic AI, like Frankenstein, but like Prometheus back in the Greek myth, they will have a question – do we give it fire? Do we give these creatures self-knowledge and self-consciousness?...And will it be similar to ours?"

Stephen Fry

That is the same challenge we face today. The implications of the choices for the future of AI are unclear. Both Geoffrey Hinton and Stephen Fry sense a danger but cannot yet fully perceive it. But the immediate benefits of AI are clear and significant. Thomas Siebel states in his book *Digital Transformation*:

"The economic benefits will be significant. McKinsey estimates that AI will increase global GDP by about $13 trillion in 2030, while a 2017 PwC study puts the figure at $15.7 trillion – a 14 percent increase in global GDP."

This makes it difficult to argue to slow down and assess the risk of AI – like Themistocles's argument to invest the silver. The scientific research on this risk is not yet available. Our capacity to assess the risk of AI is struggling to keep pace with its technological evolution.

All we have currently to point to is popular entertainment that envisions the risk of AI. Movies such as *The Matrix, The Terminator, I, Robot,* and recently *The Creator.* These movies are being produced with increasing frequency and permeate cultural consciousness. But they are not yet scientific fact – only science fiction. They cannot yet be used in a credible risk-reward analysis of AI.

We must search for the empirical facts that senior leaders typically require to make sound strategic choices. They must be scientific facts produced by acknowledged experts in their respective fields. The search for these facts is our challenge. This search must be methodical to ensure senior leaders are able to see the logical pattern. The first step is defining AI in simple terms.

2

CHAPTER 2: ARTIFICIAL INTELLIGENCE

"We're at the beginning of a golden age of AI. Recent advancements have already led to invention that previously lived in the realm of science fiction – and we've only scratched the surface of what's possible."

Jeff Bezos, Amazon CEO

2.1   WHAT IS ARTIFICIAL INTELLIGENCE (AI)?

"It is the science and engineering of making intelligent machines, especially intelligent computer programs. It is related to the similar task of using computers to understand human intelligence, but AI does not have to confine itself to methods that are biologically observable."

John McCarthy, "What is Artificial Intelligence?"

Artificial intelligence enables machines and computer programs to think critically and make decisions without direct human intervention. Tasks that once only could be performed by humans can now be done by AI. Basically, it removes the need for humans in what was previously human activity.

The tasks performed by AI range from the relatively simple to the more complex. Some AI requires the input of data by humans to learn and improve, while others can do so on their own. The technological design of

AI varies depending on the specific activity's requirement. I will repeat the most important fact again:

AI makes humans unnecessary

2.1.1 AI ALGORITHMS

"It is customary to offer a grain of comfort, in the form of a statement that some peculiarly human characteristic could never be imitated by a machine. I cannot offer any such comfort, for I believe that no such bounds can be set."

Alan Turing, The Father of Artificial Intelligence

At the core of all AI machines and computer programs is an algorithm. There would be no AI without an algorithm. Algorithms are used in mathematics and computer science. An algorithm is a procedure or set of instructions for solving a problem by calculation or other operations. An AI algorithm is a very complex algorithm. It can either be comprised of input/output data sets or a complex set of predefined rules.

These AI algorithms determine the AI's logical steps and learning ability during the problem-solving process. As a result, the AI algorithm is the program that enables the machine or computer to learn and decide independently of human intervention. It is a replacement for the human brain. I will repeat the most important fact again:

AI algorithms are a replacement for the human brain

2.1.2 MACHINE LEARNING

Machine learning is a subset of AI algorithms that utilizes input/output data sets. These data sets serve as examples and experiences by which the AI algorithm learns. Machine learning mathematically analyzes patterns across a broad variety of input data sets and draws logical inferences. A simple example is provided by Thomas Siebel in his book *Digital Transformation 2019*:

"An example of machine learning is an algorithm to analyze (input) and classify it as an "airplane" or "not airplane" (output) – potentially useful in air traffic control and aviation safety, for example. The algorithm is "trained" by giving it thousands or millions of images labeled as "airplane" or "not airplane". When sufficiently trained, the algorithm can then analyze an unlabeled image and infer with a high degree of precision whether it's an airplane."

It takes significant time and resources to develop machine learning capabilities. However, once the machine learning algorithm is honed it performs the task at light speed compared to a human being. In fact, human intervention would significantly slow down the machine.

2.1.3   DEEP LEARNING

Deep learning is a type of machine learning. It does not require extensive input/output data sets. The algorithm itself learns the patterns without human intervention. The removal of this constraint on AI learning is necessary when the data set is infinitely large. The traditional machine learning training process at that point is not feasible. A simple example provided by Thomas Siebel in his book *Digital Transformation 2019*:

"Consider, for instance, the problem of image recognition, such as creating an algorithm to recognize cars – a critical requirement for self-driving technology. The number of variants in how a car may appear is infinitely large – given all the possibilities of shape, size, color, lighting, distance, perspective, and so on. It would be impossible for a data scientist to extract all the relevant features to train an algorithm."

Deep learning AI algorithms work by dynamically building a conceptual hierarchy out of the available data set. The AI program moves across the hierarchy slowly narrowing down the possibilities of what kind of car it could be. However, the deep learning AI algorithm can be given a random car image and determine what exact car it is at machine speed. Deep learning is a form of artificial neural network technology.

2.1.4   ARTIFICIAL NEURAL NETWORKS (ANNS)

Artificial neural networks are a variation of the human brain. However, the actual structure of ANNs is very different than that of the human brain. Thomas Siebel in Digital Transformation sums up what ANNs are:

"...deep learning employs "neural network" technology...an approach originally inspired by the human brain's network of neurons, but in reality having little in common with how our brains work."

ANNs are built on the basic principles of the human brain's neural network organization. ANNs mimic the way that human neurons signal to one another in the brain. This is accomplished by a series of artificial neurons organized in layers connecting to each other. These connections are governed by a set of logical criteria (e.g., rules, weights, thresholds, etc.).

ANNs are like the human brain's neural network as they rely on informational inputs to learn and improve accuracy. But ANNs can process new information at machine speed. Google's search AI algorithm is an example of an ANN we use each day. I will repeat the most important fact again:

ANNs are an artificial variation of the human brain

2.2   A BRIEF HISTORY OF ARTIFICIAL INTELLIGENCE

"Sometimes it is the people no one can imagine anything of who do the things no one can imagine."

Alan Turing, The Father of Artificial Intelligence

The story of AI really begins with humanity's ancient philosophers and mathematicians. Foremost among these were the ancient Greeks who developed significant scientific advancements in almost every academic field. Socrates, Plato, and Aristotle enhanced humanity's capacity to reason. Thales, Pythagoras, and Archimedes developed mathematics which unlocked secrets of nature.

Formal reasoning and mathematics are at the core of AI algorithms. In large part, it is the ancient Greek's quest to uncover the secrets of the natural world that created the foundation for today's scientific thought. I will repeat the most important fact again:

The logical roots of Artificial Intelligence is Ancient Greek Thought

The next great leap in the story of artificial intelligence was in the 20th century. In 1935, Alan Turing described a computer program that could change or improve its' program. A computer program would do so by using the binary number system of "O" and "1". This enabled a computer to perform mathematical deduction and formal reasoning.

Turing's concept became known as the Universal Turing Machine or as we call it today – a computer. Warren McCulloch and Walter Pitts in 1943 described the first mathematical model of a neural network. It laid out the design for Turing-complete artificial neurons.

In 1948, another intellectual step was taken in AI research. Norbert Weiner published a book *Cybernetics: or Control and Communication in the Animal and the Machine.* Weiner compared the human brain to the computer. He speculated that computers would compete with humans in games such as chess. A machine's dynamic learning capacity would make this competition possible. In addition, Wiener thought computers would eventually replicate themselves. I am going to repeat this important fact:

Norbert Weiner thought machines would replicate themselves

In 1956, John McCarthy established the academic field of artificial intelligence. He titled a ten-week research project at Dartmouth College a "study in artificial intelligence". This was the first time the term artificial intelligence was used. In addition, what is believed to be the first AI program ever, The Logic Theorists, was presented. This AI program's purpose was to solve various math theorems from the book *Principia Mathematica*. The AI program produced a better version of a specific math theorem from the book.

The Dartmouth College research project initiated what is now known as the "Golden Age of AI" research from 1956-1974. The rapid advances in computers were a driving force for the field of AI during this period. In addition, "Cold War" competition drove heavy investment in technology (i.e., space, nuclear, military, etc.) by governments world-wide. During this period publications regarding AI (i.e., academic papers, books, etc.) significantly increased. In his book Artificial Intelligence Basics, Tom Taulli describes the competing visions for the future of AI:

"But there were generally two major theories of AI. One was led by Minsky, who said that there needed to be symbolic systems. That meant that AI should be based on traditional computer logic or preprogramming – that is, the use of approaches like If-Then-Else statements...Next, there was Frank Rosenblatt, who believed AI needed to use systems similar to the brain like neural networks...A system would be able to learn as it ingested data over time."

Minsky's vision initially won out. Neural networks were not close to being workable with existing technologies. But it only proved to be a short-lived victory. Optimism for the future value of AI technologies eventually waned. The academic community became increasingly skeptical of Minsky's vision for AI. Despite the initial promise of AI, many felt results thus far had failed lofty expectations.

Government investments in AI research began to diminish during the 1970s. This period, lasting until the early 1980s, has come to be known as the "AI winter". This led many AI researchers to abandon the field. Academics pursuing AI research deliberately omitted the term "artificial intelligence". Instead, they used terms such as machine learning or pattern recognition. However, even during this winter in AI research a beacon of light emerged.

Professor Geoffrey Hinton received his Ph.D. from the University of Edinburgh in 1972. He believed that Rosenblatt's vision of neural networks was the real future for AI research. Hinton realized that the real constraint on AI was computing power.

But Moore's Law predicted a dramatic increase in computing power in the next few decades. This would remove the primary constraint on achieving AI's neural network vision. He built on the research of others committed to a neural network vision for AI. Hinton developed new core theories in AI that today we recognize as deep learning. He published a breakthrough paper called "Learning Representations by Propagating Errors."

Hinton's breakthrough research was then followed by a succession of breakthroughs in academic research and technological innovation. AI pioneers developed neural networks based on animal brains, reinforcement learning, descent algorithms, etc. Then the logistical constraints of technology were removed with the advent of the internet, advanced search engines, graphic processing units, etc. Google was an early adopter of deep learning with the launch of the "Google Brain" project in 2011. Google then hired Hinton himself. I will restate this most important fact again:

AI's optimal solution design is an artificial variation of the human brain

Today, humans are using AI almost every day without realizing it. Google Search, YouTube, Amazon, Netflix, Siri, Alexa, and ChatGPT all utilize some form of AI. So, AI has been part of our life for quite some time. However, 21st century technological innovations have dramatically improved the value of AI. As a result, the field of AI is rapidly expanding in use and complexity.

2.3 HUMANITY'S UNCERTAIN AI FUTURE

"Success in creating AI, could be the biggest event in the history of our civilization. Or the worst. We just don't know. So we cannot know if we will be infinitely helped by AI, or ignored by it and side-lined, or conceivably destroyed by it."

Stephen Hawking, Physicist

Artificial intelligence is already a powerful force on earth that is daily increasing in power. Most AI experts agree on this fact. But what are the implications of fully unleashing that power for the first time in history? The leaders of the Manhattan Project that developed the first atomic bomb during World War 2 faced a similar question. They believed that there was a non-zero chance that igniting the first atomic bomb could destroy the entire planet. They did not yet fully understand the power of what they had created.

The risk of humanity's extinction never became a reality during the atomic bomb's testing in New Mexico. But humanity did discover the awesome power of atomic energy as a weapon. After atomic bombs were dropped on both Nagasaki and Hiroshima the power of the atom was practically understood. As a result, humanity has never used atomic power as a weapon since. It is because in an instant everyone realized that it was too powerful for use in human conflict.

Instead, humanity harnessed the power of the atom for non-military purposes such as nuclear power plants. More than half of the U.S.'s clean energy comes from nuclear power. Nuclear energy provides electricity for homes and businesses, power for desalination plants to create more fresh water, heat for refining metals, and has potential to create hydrogen as an alternative to fossil fuels.

What about artificial intelligence? Much like atomic energy our capacity to discover and use its power has outpaced our understanding. We can already see the tangible benefits of using AI. There is and should be great optimism for the application of AI to help humanity solve some of its toughest problems.

In fact, we probably cannot adequately estimate the true value of AI to humanity at this point. But what of AI's potential downside? If we as a species are going to continue investing in such a powerful new tool, we must understand the potential threat of such power.

2.4   THE EXISTENTIAL THREAT

"Humans should be worried about the threat posed by artificial intelligence."

Bill Gates, Founder of Microsoft

The threat of AI has been part of humanity's popular culture since the 20th century. Movies such as *Space Odyssey 2001*, *The Terminator*, and *The Matrix* expressed the threat of AI in different contexts. Each movie depicted a reality where AI exercised a form of free will and entered into direct conflict with humanity.

In the case of *The Terminator* and *The Matrix*, AI toppled humanity from its evolutionary apex position on earth. In the *Terminator* movies humans were seen as direct competition being hunted to extinction by AI. Conversely *The Matrix* movies humans were preserved by AI to effectively serve as livestock.

Some might think that humanity's uneasiness with artificial intelligence has been created by such modern art forms. However, humanity's fear of an artificial lifeform predates even the era of artificial intelligence itself. In 1818 Mary Shelley's *Frankenstein* grabbed hold of the public's consciousness. The main character, Dr. Victor Frankenstein, creates an artificial life form through processes other than that of natural evolution. He then loses control of the artificial organism due to mutually exclusive self-interests.

A conflict then develops between the two characters which leads to an escalating cycle of violence and death. Ultimately, Dr. Frankenstein dies attempting to destroy the artificial lifeform he created. In the story of *Frankenstein* lies humanity's subconscious fear of AI – that our own artificial creation will become our greatest rival.

But those artistic expressions of the fear of AI are science fiction. What is the real risk of AI? Humanity needs a practical, fact-based viewpoint from which to assess the risk to our species. The lens of existing risk management concepts is the best way to view this subject. In this context risk is just a degree of uncertainty in the future which cannot be 100% predicted. A positive risk is an opportunity, and a negative risk is a threat. Typically opportunities should be aggressively exploited, and threats mitigated to the maximum extent possible.

Before you can effectively mitigate a threat, you must first understand it. This is why the first step in the risk management process is to identify, define, and document the characteristics of each threat. Thus far the global community has identified the threat of AI as potentially on par with atomic weapons. But unlike with atomic weapons, humanity does not yet have sufficient clarity to define the threat in practical terms. This is due to both the complexity and speed at which AI is evolving.

So, at this point we feel the danger only instinctively. We feel a vague sense of danger akin to how we reacted to the story of *Frankenstein*. Our conscious mind cannot yet fully imagine the danger of AI and how it could become a reality. But our subconscious mind understands the danger in the most basic, primitive terms. However, we are struggling to articulate to ourselves what our instincts are trying to tell us. Therefore, it is only by discovering what AI really is in context of the earth's history that we will be able to fully understand its risk to humanity.

3

CHAPTER 3: THE DIGITAL REALITY

"The Fourth Industrial Revolution has digital technology transforming and fusing together the physical, biological, chemical, and information worlds."

> *Tony Saldana, "Why Digital Transformations Fail"*

3.1 CORE DIGITAL TECHNOLOGIES

3.1.1 THE DIGITAL BIG FOUR

"Cloud computing, big data, AI, and IoT [Internet of Things] converge to unlock business value estimated by McKinsey of up to $23 trillion annually by 2030."

> *Thomas Siebel, "Digital Transformation"*

There are four digital technologies at the core of today's digital revolution – cloud computing, big data, artificial intelligence, and the internet of things (IoT). These digital technologies are different than previous technological revolutions. Digital technologies are fundamentally changing human activity with no aspect going untouched.

The digital transformation of human activity has only begun, but already dramatic changes have occurred in warfare, business, and society. These changes will only accelerate going forward. A massive shift in competitive advantage, power, influence, and wealth is occurring across the planet. It is a sink or swim reality for all competitors in human activity. As Thomas Siebel states in his book *Digital Transformation*:

"And the stakes have never been higher – in terms of both the risk of extinction and the potential rewards."

Before we can discuss the effect of digital technologies, we must first understand what they are. We previously discussed in detail AI in chapter 2. Below are brief descriptions of the remaining three core digital technologies.

3.1.2   CLOUD COMPUTING

"Computing without Limits...The elastic cloud has effectively removed limits on the availability and capacity of computing resources – a fundamental prerequisite to building the new classes of AI and IoT applications that are powering digital transformation."

Thomas Siebel, "Digital Transformation"

Cloud computing is the foundational technology of humanity's digital transformation. Cloud environments provide a large technical ecosystem with which organizations and individuals can integrate their human activity. With this ecosystem humans have rapid internet access to a huge pool of shared information technology resources (e.g., applications, networking, data storage, etc.) with minimal management overhead. This enables organizations and individuals to move to a metered payment model which eliminates heavy up-front investments in equipment, software, etc. The cloud utilizes an elastic provisioning model. This enables organizations to flexibly scale up and scale down computing resources. As Thomas Siebel states in his book *Digital Transformation*:

"Without cloud computing, digital transformation would not be possible."

The public offering of Amazon Web Services (AWS) in 2006 forever changed human activity. AWS and similar cloud providers removed a significant constraint on human activity across the planet. Obtaining IT resources now takes only hours instead of years.

Small startups can rapidly scale out their IT resources in the cloud to compete with companies such as Microsoft, IBM, Google, etc. This removed a significant competitive barrier to human activity – warfare, business, politics, etc. The cloud's elastic nature effectively eliminates IT resource investment waste. This combination of speed and efficiency makes digital disruption possible.

3.1.3 BIG DATA

"When the data set is sufficiently complete that we can process all the data, it changes everything about the computing paradigm, enabling us to address a large class of problems that were previously unsolvable."

Thomas Siebel, "Digital Transformation"

Data is a collection of values that conveys information about some object. For example, the statement, "A man, six feet tall with light brown hair and blue eyes". A man, six feet tall, light brown hair, and blue eyes are each a separate piece of data. But without context the data isn't that meaningful. Now put it in the context of me as a person, "David is a man, six feet tall with light brown hair and blue eyes." The four combined data points with the new context of "Dave" becomes information. That information could help you pick me out of a crowd in a coffee shop. The larger the data set gets the more value can be mined from it.

For example, if you aggregated the data of men who share those four data points you may generate valuable insights. Theoretically, you might discover that we have a higher probably of car accidents on cloudless days. It might be due to the fact that blue eyes are more sensitive to sunlight. Sunglass manufacturers build entire marketing campaigns and sales strategies around this type of information. As Thomas Siebel states in his book *Digital Transformation*:

"Data, of course, have always been important. But in the era of digital transformation, their value is greater than ever before."

Collection of data across many IT ecosystems used to be expensive and time consuming. So, most organizations simply used statistical sampling to draw inferences from large data sets. But as we all know with political polling, sometimes statistical sampling does not produce accurate results. When an organization is making strategic decisions, any statistical error could prove fatal.

The advent of cloud environments has enabled the collection of massive amounts of data in a small number of ecosystems. Cloud ecosystems provide unlimited and elastic computing resources with which to analyze data. So, organizations don't need to do statistical sampling anymore. Organizations can now rapidly analyze vast amounts of source data.

So, let us use the blue eyes and sunny day accident correlation example again. A sunglass manufacturer might have taken years of expensive research by experts in data analysis to achieve that insight. Now that insight can potentially be achieved in days or even hours at much lower financial cost. That insight could lead to a new sunglass product line. A produce line that provides competitive advantage in the industry. That is the power and value of big data.

3.1.4   THE INTERNET OF THINGS (IOT)

"Inexpensive AI supercomputers the size of credit cards are deployed in cars, drones, surveillance cameras, and many other devices. This goes well beyond just embedding machine-addressable sensors across value chains: IoT is a fundamental change in the form factor of computing, bringing unprecedented computational power – and the promise of real-time AI – to every manner of device."

Thomas Siebel, "Digital Transformation"

The term IoT describes the connection of any device to the internet. The device must possess sufficient technical capabilities to; 1) connect to the internet and 2) send and receive data. Today microprocessors are getting increasingly smaller and networks much faster. As a result, more and more devices in the natural world are being connected to cyberspace. This is already pushing the full immersion of cloud computing capabilities into human activity.

Examples of devices now connected are cars, doorbell cameras, home devices, drones, industrial machinery, smartphones, laptops, and watches just to name a few. In the future other devices such as pacemakers will be connected and monitored in real-time by health professionals. In time, IoT will include so many devices a digital "mirror" will be created reflecting most of the objects we utilize in our everyday lives. All that data will be sent back in real time to the cloud to be analyzed by artificial intelligence.

This evolving dynamic driven by IoT also has the reverse effect. It is pushing the edge of human activity to be fully immersed in cyberspace – in effect blurring the line between the natural and digital worlds. This will lead to the digital reality will become more than just a data mirror of human activity.

Artificial intelligence will increasingly exercise direct control over activities in the natural world. Artificial intelligence will go beyond influencing human behavior and manipulating physical objects directly. We will eventually come full circle where humans can manipulate anything in cyberspace and artificial intelligence will be able to manipulate anything in the natural world. As Thomas Siebel states in his book *Digital Transformation*:

"The technical name for IoT – cyber-physical systems – describes the convergence and control of physical infrastructure by computers."

In effect, the digital world and the natural world will become one practical space for activity – both human and artificial. The long-term implications of this are profound.

3.2   DIGITAL TRANSFORMATION

"As mentioned earlier, digital transformation is the modern-day fight to survive the existential threat of the digital disruption caused by the Fourth Industrial Revolution."

Tony Saldana, "Why Digital Transformations Fail"

Digital transformation is about adopting digital technologies and integrating them into all aspects of an organization. It isn't about improving the current strategies, models, and outcomes produced by an organization. An industrial revolution requires a paradigm shift for the future vision of what an organization will become.

The big four digital technologies don't just enable the removal of technological constraints, but also enable the removal of imaginative constraints on strategic thinking. With successful digital transformation, previously unimaginable realms of possibilities are now available to organizations. In fact, it sometimes requires leaders to dramatically reimagine strategies and business models to seize new opportunities. Thomas Siebel states in his book *Digital Transformation*:

"In 2000, Netflix CEO Reed Hastings proposed a partnership...and Blockbuster declined. As I'm writing this, Netflix has a market capitalization of over $160 billion and Blockbuster no longer exists. Netflix saw the shift happening, discarded mail order, and transformed into a streaming video company. Blockbuster did not."

Practically speaking, digital transformation is the evolution of an organization to an entirely new way of organizing, working, and thinking. It means technically implementing digital technologies while simultaneously transforming all other aspects of the organization to integrate with those technologies. As in the Netflix example above, sometimes the outcome of digital transformation results in radical organizational change. Tony Saldana provides a detailed description of what digital transformation is in his book *Why Digital Transformations fail*:

"Digital transformation: The migration of enterprises and societies from the Third to the Fourth Industrial Revolution era. For companies, this means having digital technology become the backbone of new products and services, new ways of operation, and new business models."

3.3   DIGITAL DISRUPTION

"Most people don't realize that digital disruption is the Fourth Industrial Revolution."

Tony Saldana, "Why Digital Transformations Fail"

As Tony Saldana states above, digital disruption is really a specific and practical expression of an emerging industrial revolution. This digital revolution is being driven by rapidly evolving digital technologies. These technologies remove constraints from both organizations and competitive spaces at an ever-increasing rate. In the Digital Age, digital disruption is inevitable for every organization. Below Tony Saldana provides a detailed definition for digital disruption in his book *Why Digital Transformations Fail*:

*"**Digital Disruption**: The effect of the Fourth Industrial Revolution in the corporate and public sector landscapes. Increasingly pervasive and inexpensive digital technology is causing widespread industrial, economic, and social change. This is explosive change has occurred only in the past decade or two."*

This evolution of organizations into the Fourth Industrial Revolution is creating volatility across all human activity. Digital volatility is being created in primarily two ways; 1) self-disruption and 2) competitor disruption. These are both "volatility events" that an organization must survive to avoid extinction. Let us discuss both separately.

Self-disruption is when a thought leader in a competitive space proactively initiates a digital transformation effort. The intent is to take on the risk of digital transformation to obtain the benefit of a new competitive advantage in an existing or new competitive space.

Competitor disruption is when an organization in your existing or other competitive space digitally transforms. The new competitive advantages which that competitor gains threaten your strategic positioning. In addition, as more organizations digitally transform, this sporadic volatility in competitive spaces will only continue to increase in both frequency and intensity.

3.4   DIGITAL PUNCTUATED EQUILIBRIUM

"Darwinian evolution is a force of continuous change...By contrast, punctuated equilibrium suggests that evolution occurs as a series of bursts of evolutionary change...Today we are seeing a burst of evolutionary change – a mass extinction among corporations and a mass speciation of new kinds of companies."

Thomas Siebel, "Digital Transformation"

Some existing constraints powerfully shape the competitive dynamics in a particular competitive space. This restricts the competitive possibilities of that space. We can call these types of constraints, "keystone" constraints. Keystone constraints function like the keystone of an arch. For example, access to large amounts of computational power was such a constraint prior to the advent of cloud services.

Digital or technological punctuated equilibrium is a recurring pattern in history. Humanity has repeatedly reached the limits of its technology and then become constrained. But humanity has consistently technologically innovated and removed these limitations. The ensuing change has often led to a form of technological punctuated equilibrium. Digital punctuated equilibrium is an extreme form of digital disruption. As Thomas Siebel stated in his book Digital Transformation:

"When science and technology meet social and economic systems, you tend to see something like punctuated equilibrium. Something that has been stable for a long period suddenly disrupts radically – and then finds a new stability. Examples include the discovery of fire, the domestication of dogs, agriculture, gunpowder, the chronograph, transoceanic transportation, the Gutenberg Press, the steam engine…and the internet."

Digital disruption occurs when digital technologies remove, or partially remove, that technological keystone constraint. This causes new competitive possibilities to emerge within a competitive space (e.g., an industry, theater of war, etc.). Organizations inevitably aggressively exploit those new competitive possibilities. This leads to the development of new competitive dynamics in the competitive space. These new competitive dynamics gradually favour some organizations over others in the fullness of time. This ultimately leads to significant changes in strategic positioning within competitive spaces.

But unlike digital disruption, digital punctuated equilibrium happens, not gradually, but rapidly. In fact, it happens so rapidly it creates extreme volatility in the existing competitive dynamics. As a result, this becomes an immediate extinction threat within the competitive space. Those organizations that cannot adapt fast enough either lose significant strategic positioning or go extinct. This is a variation on how punctuated equilibrium occurs in the natural world. I'm going to repeat this important fact:

Digital punctuated equilibrium is an artificial
variation of natural punctuated equilibrium

3.5   CYBER-PHYSICAL CONVERGENCE

"Critical Cyber Trends Converging...Several critical governmental, commercial, and societal changes are converging that will threaten a safe and secure online environment. In the past several years, many aspects of life have migrated to the Internet and digital networks. These include essential government functions, industry and commerce, health care, social communication, and personal information..."

Director of National Intelligence, "Worldwide Threat Assessment (2014)"

Cyber-Physical convergence is already happening. We discussed this phenomenon earlier in the section on the Internet of Things (IoT). Others have taken notice as well. The above quote is from the U.S. Director of National Intelligence (DNI) and appears in U.S. Army's *Operational Environment and Army Learning* publication from 2014. The stated purpose of that publication is quoted below:

"Training Circular (TC) 7-102 presents concise and enduring doctrine-based guidance on how to integrate the variables of an operational environment (OE) into Army training, education, and leader development. This TC includes concepts and capabilities (products, services, and support) developed for the Army as an Operational Environment Enterprise (OEE) to improve and sustain Army readiness."

The U.S. Army is deadly serious about training and readiness. They don't discuss the implications of cyber-physical convergence in a future theoretical sense. This isn't an academic exercise to debate and theorize about. It is a fact of their combat operating environment.

To ensure mission success the U.S. Army must prepare soldiers to perform in evolving operational conditions. In effect, this is an early description of cyber-physical convergence as it was beginning to occur. This is like the description Thomas Siebel made about IoT which I repeat below:

"The technical name for IoT – cyber-physical systems – describes the convergence and control of physical infrastructure by computers."

However, Thomas Siebel was only highlighting the economic and societal opportunity of cyber-physical convergence. Another author, Bruce Schneier the former Chief Technology Officer of Resilient Systems, wrote a book that highlights the security threat posed by cyber-physical convergence. In his book *Click Here to Kill Everybody*, he describes threats of integrating cyber and physical into a single operating environment. The threat is succinctly stated by Schneier below:

" "Click here to kill everybody" is hyperbole, but we're already living in a world where computer attacks can crash cars and disable power plants – both actions that can easily result in catastrophic deaths if done at scale. Add to that hacks against airplanes, medical devices, and pretty much all our global critical infrastructure, and we've got some pretty scary scenarios to consider."

Schneier is just as deadly serious as the U.S. Army regarding the threat of cyber-physical convergence. He describes real-world examples such as hackers assuming control of a Jeep Cherokee while moving on a highway. This is a new phenomenon in history. It is a profound change in human existence and will have practical impacts on everyday life.

Cyber-physical convergence is effectively creating a new set of joint operating conditions for human activity. This is what the United States Army publication was practically describing back in 2014. This trend will only accelerate as the Fourth Industrial Revolution digitally transforms human activity. This will affect both individual organizations and entire competitive spaces such as industries and geopolitical regions. There are experts who have already identified this emerging pattern in operating conditions of human activity.

3.6 CYBER-PHYSICAL OPERATING CONDITIONS

3.6.1 UCA-S

"That's because nothing is more central to modern organizations than their capacity to cope with complexity, ambiguity, and uncertainty – in short, with spastic change."

Warren Bennis & Burt Nanus, "Leaders. Taking Charge."

There is always a "canary in a coal mine" with every emerging phenomenon. There are always Leaders with sufficient imaginative capacity to perceive something intuitively before the facts are evident. Warren Bennis and Burt Nanus are two such leaders. In their book *Leaders. Taking Charge.*, published in 1985, the authors described the conditions which are now being created by cyber-physical convergence.

The "spastic change" they described was how they were partially defining the volatility which the Fourth Industrial Revolution was beginning to create. In effect, they were describing the operating conditions which digital disruption and digital punctuated equilibrium were just beginning to create. These are the same operating conditions within which the U.S. Army is now training soldiers to operate. But this shouldn't be surprising given Warren Bennis' background.

In 1943, Warren Bennis was one of the youngest infantry officers to serve in the U.S.'s European Theater of War. He received the purple heart and bronze star for his valor. The reason Bennis understood the new operating conditions emerging in business is because he was already intimately familiar with them. He intuitively understood the pattern. It was a variation of the operating conditions in which he executed military operations on the battlefield of Europe. The operating conditions of the business world were beginning to evolve into those of the natural world. I am going to repeat this important fact:

__Business operating conditions were becoming an__
__artificial variation of natural operating conditions__

3.6.2 TUNA

"We use social ecology theory to explain and guide the effectiveness of scenario planning under what we call TUNA conditions – conditions of turbulence, uncertainty, novelty, and ambiguity – that characterize a more connected, plural, and multipolar world."

Rafael Ramirez & Angela Wilkinson, "Strategic Reframing"

Almost thirty years later, two new authors from Oxford University Rafael Ramirez and Angela Wilkinson, described the new operating conditions being created by cyber-physical convergence. Their book focuses on a necessary adaptation in strategic planning to account for the "spastic change" or volatility created by the increasingly interconnected and interdependent world. As stated in their book *Strategic Reframing*:

"While the experience of increasing change is not a new phenomenon, there is today a common perception of a quickening pace of more disruptive, large-scale changes that make the world less stable than "normal." There is thus a demand for approaches in strategic management that better prepare organizations for sudden shifts and rapidly emerging possibilities."

The authors were providing a management solution for the same pattern of operating conditions Bennis and Nanus described. Their scenario planning approach focused on the management process rather than the leadership process. But essentially, they are offering solutions to the same problem. As the challenge of cyber-physical convergence is pervasive across all elements of human activity.

3.6.3 DIGITAL VUCA

"The number and magnitude of challenging events are increasing, and the environment in which today's organizations operate is often described as volatile, uncertain, complex, and ambiguous (VUCA)."

Axelos Global Best Practice, "ITIL 4: Digital and IT Strategy"

The ITIL framework, initially published in 1989, is an IT service management framework created by the United Kingdom's Central Computer Telecommunications Agency (CCTA). Basically, it was conceived as a framework for managing IT services throughout the service lifecycle. This has evolved into a fourth version of the ITIL framework, ITIL 4, an operating model designed to meet the demands of the Digital Age.

The ITIL 4 Framework is designed to address the challenges and risks of digital transformation, digital disruption, and digital punctuated equilibrium. The authors leverage a military anacronym to describe the operating conditions of the digital age – VUCA. The anacronym VUCA stands for – volatility, uncertainty, complexity, and ambiguity. As stated in the Axelos publication *ITIL 4: Digital and IT Strategy*:

"The term 'VUCA' was coined by the US Army War College and later adopted to describe the business, social, and economic environment...The pace of change is increasing, the number of components in systems is growing, and cause-and-effect logic is becoming less linear. These changes are stimulated by digital transformation, the service economy, and other internal and external trends."

The authors of the ITIL 4 Framework perceived the same operating conditions as Bennis and Nanus, and Ramirez and Wilkinson. In fact, the entire redesign of the fourth version of ITIL is designed around operating in this emerging set of conditions. This premier IT service management publication is describing these conditions using a military concept – originally coined to describe natural world operating conditions. As with Warren Bennis being a former infantry officer describing UCA-S, this is not a coincidence. I'm going to repeat the important fact inferred from this entire section below:

> ***Natural operating conditions are emerging across human activity due to cyber-physical convergence***

3.7   AN EMERGENT PATTERN

3.7.1   A PATTERN RECOGNIZED

"I am not sure history repeats itself, but it does seem to rhyme. In management, I find one of the most important skills is pattern recognition: the ability to sort through complexity to find basic truths you recognize

from other situations. As I approach my pursuits in information technology, my decisions and choices are made in historical context."

Thomas Siebel, "Digital Transformation"

The above quoted words from Thomas Siebel's book *Digital Transformation* are the first words to appear immediately following the Preface. He begins the book by pairing two simple ideas: 1) patterns and 2) history. If I may paraphrase Mr. Siebel here, he states that he makes his decisions and choices based on the patterns of history. Why? It is because humanity has long understood that both the patterns of our history and nature repeat themselves. It is a valuable management skill to recognize these historical patterns as they are often difficult to perceive until they fully emerge into view. As the American journalist Sydney J. Harris once said:

"History repeats itself, but in such a cunning disguise that we never detect the resemblance until the damage is done."

So, patterns drive both history and nature, but they lie concealed in our everyday lives – until these patterns emerge into humanly perceptible forms. However, by the time we perceive the pattern it is already occurring, and we must react to events in real-time. If we are lucky, we recognize enough of the pattern to get ahead of the curve and to take proactive action. This normally increases the chances of success in any area of life.

Thomas Siebel, Bennis and Nanus, Ramirez and Wilkinson, and the authors of the ITIL 4: Digital and IT Strategy all partially recognized such a pattern. They collectively have partially identified the emerging pattern of cyber-physical operating conditions. This emergent pattern is part of the Fourth Industrial Revolution as described by Tony Saldana in his book *Why Digital Transformations Fail*. But to fully define this pattern we must trace it back to its root source. As we have repeatedly observed, the emerging pattern is to be found in nature.

3.7.2   NATURAL VUCA

"The environment at this level is characterized by the highest degrees of uncertainty, complexity, and ambiguity, as well as tremendous volatility (VUCA) due to the compression of time in which the leader must act."

U.S. Army War College, "Strategic Leadership Primer (1998)"

The U.S. Army began using the term "VUCA" in 1987 to describe the dramatic global changes that followed the end of the Cold War. But in truth, the Army had always thought in terms of the concepts captured in the acronym VUCA. Armies throughout history have always had to operate in the natural operating conditions of land, sea, and air. The newly coined term VUCA was just a modern expression of this ancient truth. The key to understanding why this is so lies in the answer to what the root of warfare itself is.

Warfare is the singular human activity which returns humanity to its most primitive and violent patterns of behavior. However sophisticated technologies become, the patterns of war essentially remain unchanged. That is why military commanders across the planet are studying Sun Tzu's book *The Art of War* – a book written over 2,000 years ago in ancient China. In the introduction to *On War* by Carl von Clausewitz, COL. F.N. Maude succinctly states why the patterns of war repeat themselves:

"What Darwin accomplished for Biology generally Clausewitz did for the Life-History of Nations nearly a half a century before him, for both have proved the existence of the same law in each case, viz. "The survival of the fittest" – the "fittest"...as emanating from a force inherent in all living organisms which can only be mastered by understanding its nature."

Warfare is the survival of the fittest contest for our species, homo sapiens. So, the natural operating conditions emerging today across human activity are evolutionary operating conditions – such as natural punctuated equilibrium.

3.7.3   NATURAL PUNCTUATED EQUILIBRIUM

"Evolution is a theory of organic change, but it does not imply, as many people assume, that ceaseless flux is the irreducible state of nature… Change is more often a rapid transition between stable states than a continuous transformation at slow and steady rates."

Stephen Jay Gould, "The Panda's Thumb"

In 1972, paleontologists Stephen Jay Gould and Niles Eldredge published a scientific paper called *Punctuated Equilibria*. Gould and Eldredge presented a new theory within evolutionary science. Charles Darwin's theory of evolution presumed that the process of evolution generally preceded gradually over time. In contrast, Gould and Eldredge theorized that the evolution of species would remain relatively unchanged for long periods called stasis. This period of stasis would then be followed by a period of extreme evolutionary volatility. This would cause rapid populational growth of some species while simultaneously causing the extinction of other species.

Once the period of extreme volatility ended, another long period of relative statis would commence. Then Darwin's gradual evolutionary process would start again. Gould and Eldredge theorized the pattern repeated during earth's history. They called it punctuated equilibrium. The environmental conditions they described are eerily like those of the Digital Age. Thomas Siebel selected natural punctuated equilibrium as the conceptual basis for his book *Digital Transformation*. He applied this concept to examples of extreme digital disruption in human activity. His selection of this concept is consistent with the pattern identified by Bennis and Nanus, Ramirez and Wilkinson, and the authors of ITIL 4 Framework.

The difference is Thomas Siebel explicitly selected and expressed a concept in natural evolution itself. But in practical terms all these authors were partially identifying the emerging pattern of cyber-physical converge in human activity. And this is being caused by rapid changes as part of the Fourth Industrial Revolution – a pattern that is a variation of natural evolutionary conditions. This means to find the root cause of cyber-physical convergence conditions we must look further into natural evolutionary theory. I am going to repeat the important fact of this section:

> **_Digital operating conditions are an artificial variation of natural evolutionary conditions_**

4

CHAPTER 4: THE THEORY OF NATURAL EVOLUTION

"Let me lay my cards on the table. If I were to give an award for the single best idea anyone has ever had, I'd give it to Darwin, ahead of Newton and Einstein and everyone else."

Daniel Dennett, "Darwin's Dangerous Idea:
Evolution and the Meanings of Life"

4.1   NATURAL EVOLUTIONARY THEORISTS

4.1.1   CHARLES DARWIN

"No other field of science is as burdened by its past as is evolutionary biology...The discipline of evolutionary biology can be defined to a large degree as the ongoing attempt of Darwin's intellectual descendants to come to terms with his overwhelming influence."

John Horgan, "The End of Science"

Charles Robert Darwin (12 February 1809 – 19 April 1882) was an English born naturalist, geologist, and biologist. He was the first person to fully express the theory of natural evolution. It is one of the most important ideas in history.

In his book, *On the Origin of Species,* Charles Darwin conceived the idea of natural evolution occurring via the mechanism of natural selection. He theorized that the mechanism of natural selection drove the recurrence of the visible patterns we see in nature across geological time. However, he did not invent the process of natural evolution – he discovered it. Natural evolution had been occurring for billions of years prior to Charles Darwin's discovery.

4.1.2   ERASMUS DARWIN

"Would it be too bold to imagine, that in the great length of time, since the earth began to exist, perhaps millions of ages before the commencement of the history of mankind, would it be too bold to imagine, that all warm-blooded animals have arisen from one living filament, which THE GREAT FIRST CAUSE endued with animality, with the power of acquiring new parts, attended with new propensities, directed by irritations, sensations, volitions, and associations; and thus possessing the faculty of continuing to improve by its own inherent activity, and of delivering down those improvements by generation to its posterity, world without end!."

Erasmus Darwin, "Zoonomia"

Dr. Erasmus Robert Darwin (12 December 1731 – 18 April 1802) was an English born physician, natural philosopher, naturalist, physiologist, inventor, and poet. He is also the grandfather of Charles Darwin. However, he died before Charles was born.

Erasmus Darwin would today be considered a polymath – a person of wide-ranging knowledge or learning. He was one of the key English thinkers during the Age of Enlightenment. Erasmus Darwin was also one of the first to begin perceiving the pattern of natural evolution.

The excerpt above is from his book, "Zoonomia". The quote expresses a good deal of the eventual theory of natural evolution discovered by Charles. In his other writings, Erasmus Darwin identified many other foundational concepts that make up the pattern of natural evolution.

4.1.3   ON THE ORIGIN OF SPECIES

"...his mother persuaded to take along a copy of the Origin of Species, saying "It sets the door of the universe ajar!"...Sherrington's mother was right. No other scientific theory has had such a tremendous impact on our understanding of the world and of ourselves as has the theory Charles Darwin presented in that book."

Timothy Shanahan, "The Evolution of Darwinism"

Charles Darwin's book *On the Origin of Species* was published on November 24th, 1859. In this book Darwin presented his theory of natural evolution. The response to Darwin's theory of natural evolution caused conflict and debate within the scientific community. However, in time his idea won the intellectual "struggle for existence". The gradual natural adaptation of species through the process of natural selection became the intellectual foundation for evolutionary biology.

Darwin's ideas served as the intellectual starting point of reference for natural evolutionary science. In a way Darwin's conceptual framework became the intellectual context for all future natural evolutionary scientific thought. This is why Darwin's influence on the field of natural evolutionary science is so powerful even to this day.

4.1.4   RICHARD DAWKINS

"The selfish gene theory is Darwin's [Charles] theory, expressed in a way that Darwin did not choose but whose aptness, I should like to think, he would instantly have recognized and delighted in...Rather than focus on the individual organism, it takes a gene's eye view of nature. It is a different way of seeing, not a different theory."

Richard Dawkins, "The Selfish Gene"

Richard Dawkins (26 March 1941) is a British evolutionary biologist and bestselling author. He held a professorship for the Public Understanding of Science at the University of Oxford from 1995 to 2008. Dawkins published the book *The Selfish Gene* in 1976.

Since then he has published numerous books on natural evolutionary theory (e.g., *The Blind Watchmaker*, *The Extended Phenotype*, etc.). He is well known for his theory that argues for a gene-centered view of natural evolution. In this view Dawkins argues for the gene as the principal unit of selection in natural evolution.

4.1.5   THE SELFISH GENE

"The argument of this book is that we, and all other animals, are machines created by our genes. Like successful Chicago gangsters, our genes have survived, in some cases for millions of years, in a highly competitive world."

Richard Dawkins, "The Selfish Gene"

In 1976 Dawkins published the book *The Selfish Gene*. In this book he presented a new theory on a specific aspect of natural evolutionary science – genes. Dawkins theorized that each individual gene had an inherent ruthlessness in ensuring its own survival. It is this inherent ruthlessness that is the driving force of the process of natural selection.

Dawkins theorized that the gene is the centerpiece of the processes of natural inheritance, natural selection, and natural evolution itself. By reframing existing natural evolutionary thought, Dawkins enabled humanity to derive new, and potentially meaningful insights. In *The Selfish Gene*, Dawkins uses simple metaphors to explain complicated natural evolutionary concepts.

4.2   THE THEORY OF NATURAL EVOLUTION

"Thus, from the war of nature...the production of higher animals, directly follow. There is grandeur in this view of life, with its several powers, having been originally breathed by the Creator into a few forms or into one; and that, whilst this planet has gone cycling on according to the fixed law of gravity, from so simple a beginning endless forms most beautiful and most wonderful have been, and are being evolved'."

Charles Darwin "On the Origin of Species"

The theory of natural evolution revealed the underlying pattern in nature. The key mechanism of this underlying pattern was natural selection. The theory brought clarity to humanity's understanding of the natural world. It allowed us to understand how organisms adapted, evolved, speciated, expanded, and went extinct. The theory of natural evolution can generally be expressed in the following simple statements:

- Natural organisms vary in their structure and behaviors.
- Some variations are random while others are nonrandom.
- Natural organisms with beneficial variations tend to exist longer than others.
- An organism existing longer enables it to reproduce more offspring with the same variations.
- Increased offspring production generally increases an organism's population size.
- A larger population size of an organism provides advantages that lower the risk of extinction.
- The organism then has a greater chance of being naturally selected to survive and reproduce.

During this process of natural inheritance, an organism's offspring are formed within the existing constraints of the species (i.e., form, genetic information, etc.). It is like an architect adding new additions to an existing house (i.e., second floor, new rooms, etc.). He must work within the constraints of the existing foundations and structure. This limits his options in designing and building new additions to the house. It is the same in natural evolution.

Each species is developing within a set of constraints that limit the species' natural evolutionary possibilities. Think about how hard it would be for an elephant to evolve into a mouse? I'm not so sure an elephant could evolve to become so small. This constraint tends to set a rough trajectory for each species' structural and behavioral development. That's what natural evolution is – the gradual progression of a species' development along its uniquely constrained evolutionary trajectory.

That's it. The power of Darwin's theory lies in its simplicity. What could be simpler than an organism's drive to survive? That makes the definition of "winning" for a species in natural evolution to just exist and reproduce for as long as possible. The "big win" is to never go extinct. Very few species have avoided extinction. Scientists think over 5 billion species have existed during earth's history. Over 99% of all those species have gone extinct. So, very few species have achieved the "big win" in earth's evolutionary history.

Natural evolution is happening every moment of every day everywhere on earth. The pattern of natural evolution has been repeated on earth for billions of years. This pattern will continue to repeat for as long as there is life on earth.

5

CHAPTER 5: NATURAL EVOLUTIONARY CONDITIONS

5.1   NATURAL ECOSYSTEMS

"Let it also be borne in mind how infinitely complex and close-fitting are the mutual relations of all organic beings to each other and to their physical conditions of life; and consequently what infinitely varied diversities of structure might be of use to each being under changing conditions of life."

Charles Darwin, "On the Origin of Species"

Natural ecosystems are a complex web of interdependencies. Patterns of natural coevolution, natural co-adaptation, and the environment create this web. Each thread of the web possesses many-to-many relationships. This creates a set of tangled relationships between natural organisms and the environment.

In addition, this web is not static, but a moving target due to dynamic change. Darwin repeatedly pointed out this fact. He lamented this made drawing clear cause and effect relationships in nature often extremely difficult, if not impossible. This can often lead to seemingly insignificant changes in a natural ecosystem, causing a very disruptive event.

5.2 NATURAL EVOLUTIONARY CONDITIONS

"We can, in short, see why nature is prodigal in variety, though niggard in innovation. But why this should be a law of nature if each species has been independently created no man can explain...Many other facts are, as it seem to me, explicable on this theory...these facts cease to be strange, or might even have been anticipated."

Charles Darwin, "On the Origin of Species"

Each individual organism must struggle to survive in recurring natural evolutionary conditions. An extreme example is the asteroid that struck the Earth 65 million years ago. The planet's environment changed rapidly and dramatically. The organisms we call dinosaurs were unable to change fast enough. Therefore, the dinosaurs went extinct. However, other organisms such as small mammals were able to survive – our distant ancestors. The mammals did so within generally the same pattern of natural evolutionary conditions.

Natural evolutionary conditions are the factors (i.e., climate, water, geology, etc.) that shape natural ecosystems. Changes in climate can dramatically increase or decrease existing food resources. In addition, geological events such as continental drift, earthquakes, volcanic eruptions, formation of mountains, etc. can powerfully impact natural organisms. These conditions are the context within which all natural organisms must operate in the struggle for existence.

Below is a set of natural evolutionary conditions compiled from Charles Darwin's book *On the Origin of Species*. The set of conditions is a variation of the set conditions produced in Digital Reality when the anacronyms UCA-S, TUNA, and VUCA are combined.

5.2.1 AMBIGUOUS

"We can dimly see why the competition should be most severe between allied forms, which fill nearly the same place in the economy of nature; but probably in no one case could we precisely say why one species has been victorious over another in the great battle of life."

Charles Darwin, "On the Origin of Species"

Natural ecosystems are generally very complex. This complexity is due in large part to interspecies competition and environmental changes. Therefore, precisely determining cause and effect relationships within natural ecosystems is often not practically possible. As a result, events occurring in the natural world are often perceived as ambiguous by human beings.

Complex and tangled ecosystems make cause and effect relationships unclear

5.2.2 CHAOTIC

"Throw up a hand full of feathers, and all fall to the ground according to definite laws; but how simple is the problem where each shall fall compared to that of the action and reaction of the innumerable plants and animals which have determined, in the course of centuries, the proportional numbers and kinds of trees now growing on the old indian ruins!"

Charles Darwin, "On the Origin of Species"

Natural ecosystems are generally very complex. This complexity is due in large part to interspecies competition and environmental changes. Consequently, the course of events within natural ecosystems can be highly sensitive to initial conditions. Small differences in the initial conditions can cause widely diverging outcomes. As a result, human beings often perceive events occurring in the natural world to be chaotic and random.

Ecosystem events can seem random due to high sensitivity to small changes

5.2.3 COMPLEX

"Let is also be borne in mind how infinitely complex and close-fitting are the mutual relations of all organic beings to each other and to their physical conditions of life; and consequently what infinitely varied diversities of structure might be of use to each being under changing conditions of life."

Charles Darwin, "On the Origin of Species"

Natural ecosystems are generally very complex. This complexity is due in large part to interspecies competition and environmental changes. Often it is practically not possible to map all the interrelationships of the natural world – especially between organic and inorganic elements.

Changing natural conditions, environmental and competitive, creates infinite complexity

5.2.4   DYNAMIC

"Natural Selection acts exclusively by the preservation and accumulation of variations, which beneficial under the organic and inorganic conditions to which each creature is exposed at all periods of life. The ultimate result is that each creature tends to become more and more improved in relation to its conditions. This improvement inevitably leads to the gradual advancement of the organization of the greater number of living beings throughout the world."

Charles Darwin, "On the Origin of Species"

Natural ecosystems are inhabited by species that are continually changing. This change is in large part driven by changes in environmental conditions and fierce competition with other organic beings. As a result, the natural world is perceived by human beings as a fluid and dynamic space.

Ecosystems are continuously changing due to competition and environmental fluctuations

5.2.5   EMERGENT

"...a variety when once formed must again, perhaps after a long interval of time, vary or present individual differences of the same favourable nature as before...Seeing that individual differences of the same kind perpetually recur."

Charles Darwin, "On the Origin of Species"

Natural ecosystems are characterized by a pattern of recurring environmental conditions and competitive dynamics. In the span of billions of years this leads to organic life emerging which inevitably follows similar patterns. This is in part due to the combination of genetic and structural (i.e., phenotypic) constraints previously discussed. The repetitive environmental conditions are created by the immutable laws of the physical world (e.g., law of gravity, etc.). So, what human beings perceive as novel phenomena in the natural world are, in the most meaningful sense, just a reemergence.

Recurring evolutionary patterns reemerge under similar conditions and circumstances

5.2.6 EXPANSIVE

"In looking at Nature, it is most necessary to keep the foregoing considerations always in mind - never to forget that every single organic being may be said to be striving to the utmost to increase in numbers; that each lives by a struggle at some period of its life; that heavy destruction inevitably falls either on the young or old, during each generation or at recurrent intervals. Lighten any check, mitigate the destruction ever so little, and the number of the species will almost instantaneously increase to any amount."

Charles Darwin, "On the Origin of Species"

Natural ecosystems are spaces where inhabitants engage in continuous competition. The purpose of this competition for each organic being is to increase its population. This increase usually requires the geographical expansion of the organic being's range. An expanded range enables organisms to secure access to scarce resources. As a result, the organic beings on earth are perceived to be continually striving to expand, in terms of numbers and territory, unless checked by some external constraint (i.e., geographical, resource, climate, competitive, etc.).

Species are continuously competing for resources and territory in-order-to expand population size

5.2.7 PARADOXICAL

"It is a strange result which we thus arrive at, namely that characters of slight vital importance to the species, are the most important to the systematist; but, as we shall hereafter see when we treat of the genetic principle of classification, this is by no means so paradoxical."

Charles Darwin, "On the Origin of Species"

Natural ecosystems are spaces where many natural laws and patterns are constantly recurring. They are often entangled in unseen ways. As a result, natural ecosystems often present phenomenon that are paradoxical on their face.

A great example of this are the words of French writer Jean-Baptiste Alphonse Karr who wrote: "the more things change, the more they stay the same." Although the statement is contradictory, we intuitively know the statement to be true. It is the same in the natural world. However, the paradoxical nature of a phenomenon only exists in our perception. As we untangle the web of natural laws and patterns at work, the phenomenon loses its paradoxical nature.

The tangled nature of evolutionary conditions presents seemingly paradoxical phenomena

5.2.8 TANGLED

"We can clearly see how it is that all living and extinct forms can be grouped together within a few great classes; and how the several members of each class are connected together by the most complex and radiating lines of affinities. We shall never, probably, disentangle the inextricable web of affinities between members of any one class; but when we have a distinct object in view, and do not look to some unknow plan of creation, we may hope to make sure but slow progress."

Charles Darwin, "On the Origin of Species"

Natural ecosystems are bound in an inextricable web genetically, behaviorally, and, in many cases, physically. These entanglements make it hard to discern cause-and-effect relationships. As a result, the natural world is perceived by human beings as a tangled space. Often the individual significance of two or more elements can only be understood together.

__Ecosystems are a tangled web of interrelationships between species and their environment__

5.2.9 TURBULENT

"All that we can do, is to keep steadily in mind that each organic being is striving to increase in a geometrical ratio; that each at some period of its life, during some season of the year, during each generation or at intervals, has to struggle for life and to suffer great destruction."

Charles Darwin, "On the Origin of Species"

Natural ecosystems are characterized by continuous competition between their inhabitants. In many cases, species prey on and/or serve as prey for other species. In other cases, they engage in conflict over scarce resources they both require to survive and/or reproduce. As a result, the natural world is perceived by homo sapiens as a space of conflict, disorder, and violence.

__Ecosystems are places of continuous conflict between species in the struggle for existence__

5.2.10 UNCERTAIN

"No fixed law seems to determine the length of time during which any single species or any single genus endures."

Charles Darwin, "On the Origin of Species"

Natural ecosystems exist in a general state of uncertainty. The unpredictability of weather patterns is a simple example of this uncertainty. As a result, human beings perceive the natural world as an uncertain place where future events often cannot be accurately predicted.

__Ecosystems are uncertain due to the infinite number of environmental variables__

5.2.11 VOLATILE

"It is, however, probable...that the world...subjected to more rapid and violent changes in its physical conditions...and such changes would have tended to induce changes at a corresponding rate in the organisms."

Charles Darwin, "On the Origin of Species"

Natural ecosystems are spaces of constant change. In addition, these ecosystems can be very complex. These factors, combined with others, can lead to rapid and violent changes – often driven by seemingly minor changes. As a result, events occurring in the natural world can be perceived as volatile by human beings.

Ecosystems are volatile due to the potential for unpredictable, sudden, and rapid change

5.3   NATURAL CHECKS

"The causes which check the natural tendency of each species to increase... the amount of food for each species...seasons of extreme cold or drought... constantly suffering enormous destruction...from enemies or from competitors for the same place and food..."

Charles Darwin, "On the Origin of Species"

All organisms are driven to expand their population and territory. They will continue to expand unless stopped or checked by a constraint of some kind. It is a close approximation to Sir Isaac Newton's First Law of Motion which is roughly expressed as, "An object in motion will stay in motion unless acted upon by an equal or opposite force." That is what a "check" is in natural evolutionary theory. It is a force (i.e., climatic, geological, competitive, etc.) that stops an organisms' expansion or motion.

Examples of evolutionary checks are mountain ranges, islands, oceans, artic tundra, deserts, predators, droughts, epidemics, etc. A powerful example of a check in history are the black plagues. These epidemics dramatically reduced populations. This impacted the expansion of humanity for generations. Such is the awesome power of natural evolutionary checks.

5.3.1   CONSTRAINT-BARRIERS

"We can thus understand the high importance of barriers, whether of land or water, in not only separating, but in apparently forming the several zoological and botanical provinces."

Charles Darwin, "On the Origin of Species"

Natural ecosystem boundaries are often shaped by geographical features such as oceans, rivers, lakes, wetlands, deserts, tundra, mountain ranges, canyons, islands, etc. These geographic features typically act as a physical check to the migration of natural species. This has the effect of physically separating two or more populations of organic life. As a result, geographic features can be barriers to natural interbreeding and/or competition.

Geographical barriers create boundaries between populations

5.3.2   CONSTRAINT-CLIMATE

"Change of climate must have had a powerful influence on migration. A region now impassable to certain organisms form the nature of its climate, might have been a high road for migration, when the climate was different."

Charles Darwin, "On the Origin of Species"

Natural ecosystems are shaped by the long-term weather patterns or climate. The climate is the average and variance of meteorological variables (i.e., temperature, humidity, wind, rain, precipitation, etc.). Weather patterns typically act as a climatical check on the migration of natural species. This has the effect of separating two or more populations of organic life. As a result, climate can thus be seen as a barrier to natural interbreeding and/or competition.

Climate creates boundaries between populations

5.3.3   CONSTRAINT-COMPETITIVE

"Still less is it meant, that species which have the capacity of crossing barriers and ranging widely, as in the case of certain powerfully-winged birds, will necessarily range widely; for we should never forget that to range widely implies not only the power of crossing barriers, but the more important power of being victorious in distant lands in the struggle for life with foreign associates."

Charles Darwin, "On the Origin of Species"

Natural ecosystems are spaces of fierce competition between inhabitants. Competition among organic beings affects all life on earth. This competition between species acts as an invisible check on the migration of most natural species. Examples of this dynamic are predator-prey relationships, epidemics, consumption of scarce resources, etc. As a result, competition can be a constraint to the expansion of most natural species on earth.

Fierce competition between species checks their expansion

5.3.4   CONSTRAINT-PHENOTYPIC

"No doubt, as Mr. Wallace has remarked with much truth, a limit will be at last reached. For instance, there must be a limit to the fleetness of any terrestrial animal, as this will be determined by the friction to be overcome, the weight of body to be carried, and the power of contraction in the muscular fibers."

Charles Darwin, "On the Origin of Species"

Natural ecosystem dynamics drive the evolutionary advancement of organic life. Each structural adaptation by an organic being offers it a new set of competitive opportunities. However, the same adaptation also simultaneously restricts other possible adaptive trajectories (i.e., opportunity cost) over time. An elephant cannot adapt to become a mouse in a time compressed competitive window. As a result, the natural world is perceived by human beings as a place where cost benefit tradeoffs occur as a natural result of adaptation.

Structure has the potential to limit evolutionary possibilities

5.3.5   CONSTRAINT-PHYSICAL LAW

"So again it is difficult to avoid personifying the word Nature; but I mean by Nature, only the aggregate action and product of many natural laws, and by laws the sequence of events as ascertained by us."

Charles Darwin, "On the Origin of Species"

Natural ecosystems are governed by underlying natural laws, such as the law of gravity, that govern the universe. These natural laws interact with each other in a consistent pattern. They shape evolutionary patterns and affect daily events in all natural ecosystems. As a result, the natural world is perceived by human beings as occurring in repetitive patterns such as the changes in the environment (i.e., seasons, tides, etc.) and competition (i.e., speciate, vary, expand, contract, extinct, etc.).

Natural law governs the environment and competition in ecosystems

5.3.6   CONSTRAINT-RESOURCES

"The amount of food for each species of course gives the extreme limit to which each can increase..."

Charles Darwin, "On the Origin of Species"

Natural ecosystems offer limited resources to sustain organic beings. This creates a condition of resource scarcity relative to the number of organic beings. It is a driving force of organic competition. Organic beings typically produce more offspring (i.e., seek to expand) than can be supported. As a result, the natural world is perceived by human beings as being scarce in resources.

Resource scarcity is a constraint on expansion

5.3.7   CONSTRAINT-TIME

"Though Nature grants long periods of time for the work of natural selection, she does not grant an indefinite period; for as all organic beings are striving to seize on each place in the economy of nature, if any one species does not become modified and improved in a corresponding degree with its competitors, it will be exterminated."

Charles Darwin, "On the Origin of Species"

Natural ecosystems evolve over time due to environmental changes and competition. There is ultimately a time window within which organic life must adapt to survive. If organic life cannot change fast enough, then it will inevitably go extinct. As a result, the natural world is perceived by human beings as a space with compressed time windows for action (i.e., seasonal farming, reproduction, etc.).

Species have a finite time window to adapt or go extinct

5.3.8   CONSTRAINT-PERCEPTION

"But the chief cause of our natural unwillingness to admit that one species has given birth to clear and distinct species, is that we are always slow in admitting great changes of which we do not see the steps."

Charles Darwin, "On the Origin of Species"

Natural ecosystems evolve over time due to environmental changes and competition. However, the causes, effects, and sequential steps of this evolutionary process are often obscure. In addition, the perception of organic beings is limited by physical and cognitive abilities. As a result, human beings often inaccurately perceive the natural world due to visual and imaginative constraints.

Physical and cognitive limitations restrict perceptive ability

5.4   CONCLUSION

The recurring conditions and constraints of earth shape the patterns of natural evolution. In his book *On the Origin of Species* Charles Darwin described this overall pattern of natural evolution. We will next discuss the pattern he discovered.

6

CHAPTER 6: NATURAL EVOLUTIONARY PATTERNS

"This is one of the most interesting departments of natural history, and may almost be said to be its very soul. What can be more curious than that the hand of a man, formed for grasping, that of a mole for digging, the leg of a horse, the paddle of a porpoise, and the wing of a bat, should all be constructed on the same pattern..."

Charles Darwin, "On the Origin of Species"

6.1   NATURAL EVOLUTIONARY SPACES

"It is universally admitted, that in most cases the area inhabited by a species is continuous; and that when a plant or animal inhabits two points so distant from each other, or with an interval of such a nature, that the space could not have been easily passed over by migration, the fact is given as something remarkable and exceptional."

Charles Darwin, "On the Origin of Species"

Natural ecosystems are spaces inhabited by a diverse set of species. These spaces have a strong influence on the evolutionary trajectory of many species. Open spaces enable the spread of species into many different sub ecosystems. So, species adapt to these new habitats.

Confined spaces have the opposite effect – driving species to develop specialized adaptations. It is like the physical space (e.g., house design, street type, town layout, etc.) of your childhood. You developed specialized adaptations to your unique circumstances. It is the same in nature. We will discuss the spaces highlighted in *On the Origin of Species* in the rest of the section below.

6.1.1 COMPETITIVE SPACE

"...the course of modification will generally have been more rapid on large areas; and what is more important, that the new forms produced on large areas, which already have been victorious over many competitors, will be those that will spread most widely...They will thus play a more important part in the changing of the organic world."

Charles Darwin, "On the Origin of Species"

Natural ecosystems are inhabited by a diverse set of natural species. Inevitably competition will occur over scarce resources such as food, territory, etc. This competition must occur in a physical space within a natural ecosystem. When competition occurs this physical space becomes a new competitive space. A competitive space is a physical space where species are actively competing whether indirectly or directly. An example of a competitive space would be the African savannah where lions, cheetahs, and leopards compete for similar prey, and kill each other's offspring, if possible.

A competitive space is where competition actively occurs

6.1.2 CONFINED SPACE

"Thus it will be under nature; for within a confined area, with some place in the natural polity not perfectly occupied, all the individuals varying in the right direction, though in different degrees, will tend to be preserved."

Charles Darwin, "On the Origin of Species"

Natural ecosystem boundaries are often shaped by geographic features such as oceans, rivers, lakes, wetlands, deserts, tundra, mountain ranges, canyons, islands, etc. These typically check the migration of natural species. A space which is not open to migration patterns of natural species is a confined or isolated space. Examples of confined spaces would be a remote island or a land locked lake.

> **_A confined space is surrounded by barriers that restrict migration_**

6.1.3   DOMINANCE SPACE

"But when we bear in mind that almost every species, even in its metropolis, would increase immeasurably in numbers, were it not for other competing species;... in short, that each organic being is either directly or indirectly related in the most important manner to other organic beings, - we see that the range of the inhabitants of any country...depends...on the presence of other species...with which it comes into competition...the range of any one species, depending as it does on the range of others, will tend to be sharply defined."

Charles Darwin, "On the Origin of Species"

Natural ecosystems are inhabited by a diverse set of natural species. Competition drives the expansion of some natural species and the contraction of others. It is a zero-sum game given the scarcity of resources in terms of food and territory. Those that expand the most relative to their competitors will have achieved dominance – a dominance space. It is a space where a dominant natural species exerts control and influence over other natural species limiting their expansion.

> **_A dominance space is where one or more species_**
> **_exert control and influence over others_**

6.1.4   NEUTRAL SPACE

"In looking at species as they are now distributed over a wide area, we generally find them tolerably numerous over a large territory, then becoming somewhat abruptly rarer and rarer on the confines, and finally

disappearing. Hence the neutral territory, between the two representative species is generally narrow in comparison with the territory proper to each."

Charles Darwin, "On the Origin of Species"

Natural ecosystems are spaces of fierce competition over scarce resources. Often two natural species which are in direct competition will each establish mutually exclusive dominance spaces. Often there will be a buffer zone between these two natural species where neither natural species exerts control or influence over the other. This physical space where neither of the two dominant natural species exists in large numbers, we will call a neutral space. Often this neutral space is created by some environmental (i.e., climate) or geographic (i.e., mountains, etc.) barrier.

A neutral space is a physical space where there is no clear dominant natural species

6.1.5   OPEN SPACE

"Although isolation is of great importance in the production of new species, on the whole I am inclined to believe that largeness of area is still more important, especially for the production of species which shall prove capable of enduring for a long period, and spreading widely."

Charles Darwin, "On the Origin of Species"

Natural ecosystems vary in geographical size. Those natural ecosystems which are open to species migration (i.e., no barriers such as mountains, oceans, climate, etc.) are considered an open space. In open spaces competition is more dynamic and intense. Natural species that rise to dominance in open spaces tend to have the most competitive adaptations. This is due to the degree of competition with other natural species.

Open space has no significant barriers to natural species migration and competition

6.2 NATURAL ORGANISMS

"Organism – An organized being, whether plant or animal."

Charles Darwin, "On the Origin of Species"

There are many types of natural organisms on the earth. Every natural organism is made of one or more cells (the basic unit of all organic life forms). Examples of natural organisms are animals, plants, fungi, bacteria, etc. Most natural organisms must consume some resource (i.e., food, water, sunlight, etc.) to sustain a system of organs. All natural organisms reproduce, grow, develop, and maintain a system of organs. In addition, natural organisms can respond to environmental stimuli.

The word organism is derived from two Ancient Greek words. The first, 'organon', meaning instrument, tool, organ. The second, 'ismus' (a word evolved thru Ancient Greek, Latin, and finally English) roughly means a system or condition. So, an organism is, in its most basic sense, a completed system of organs. However, only some organs can be perceived by the homo sapiens eye.

We observe organisms repeating certain patterns of instinctual behavior. This complete set of visually observable traits of structure and behavior is called an organism's phenotype. An organism's phenotype is the result of influences deriving from a combination of inheritable and environmental factors. As a result, the same patterns of phenotypic forms tend to repeat in nature such as animals with four legs (e.g., triceratops, crocodiles, mammoths, rabbits, grizzly bears, etc.).

6.3 NATURAL SPECIES

"From these remarks it will be seen that I look at the term "species" as one arbitrarily given, for the sake of convenience, to a set of individuals closely resembling each other..."

Charles Darwin, "On the Origin of Species"

The concept of natural species in evolutionary theory roughly defines a large group of similar natural organisms. These natural organisms are grouped together for two major reasons. First, the group of natural organisms typically possess the same visually observable physical characteristics or phenotype. Second, this large group is capable of successfully mating and producing fertile offspring.

The species concept provides evolutionary theory a rough working model for classifying organisms. This classification aids homo sapiens in organizing and simplifying the complexity of nature. Examples of natural species are the lion, giant panda, Indian elephant, red kangaroo, jaguar, grizzly bear, and great white shark.

6.4 NATURAL EXPANSION

"In looking at Nature, it is most necessary to keep the foregoing considerations always in mind – never to forget that every single organic being may be said to be striving to the utmost to increase in numbers..."

Charles Darwin, "On the Origin of Species"

Natural organisms are instinctually driven to expand their population size. Why? It is the risk of extinction. A large population size decreases extinction risk and a small population size increases the risk. Basically, population expansion is a risk reduction strategy. Therefore, population expansion is a necessity for all organisms in the struggle for existence.

Population expansion is accomplished through the production of natural offspring or reproduction. The drive to reproduce is an instinctual behavior all organisms are born with. Each natural species has a specific reproductive strategy. This strategy is designed to increase the chances for population expansion. Organisms will expand populationally and geographically unless stopped by some external force. This fact is a key factor in natural evolutionary competition.

6.5 NATURAL OFFSPRING

"Nevertheless these cases are only exaggerations of the common fact that the female produces offspring of two sexes which sometimes differ from each other in a wonderful manner."

Charles Darwin, "On the Origin of Species"

Natural organisms reproduce themselves in the form of new natural organism(s). These new organisms are referred to as natural offspring. Natural offspring are created by one or two natural organisms during the natural reproductive process.

The natural offspring's parent(s) pass on information to the offspring. This information is a set of instructions. The instructions enable the offspring to develop the completed system of organs specific to that type of organism. This includes the information for developing its ancestors accumulated evolutionary advantages. Typically this includes the parent's same structure (form) and innate behaviors (instincts) or phenotype. In addition, this information will be leveraged by the natural offspring to produce future evolutionary advantages.

6.6   THE STRUGGLE FOR EXISTENCE

"Nothing is easier than to admit in words the truth of the universal struggle for life, or more difficult – at least I have found it so – than constantly to bear this conclusion in mind. Yet unless it be thoroughly engrained in the mind, the whole economy of nature, with every fact on distribution, rarity, abundance, extinction, and variation, will be dimly seen or quite misunderstood."

Charles Darwin, "On the Origin of Species"

Natural organisms on the earth are locked in a constant 'struggle for existence'. This is true for all life on earth. As Darwin stated, that fact is the starting point for understanding natural evolution.

Human sports have a similar starting point as well. As the National Football League (NFL) Pro Football Hall of Fame coach, Vince Lombardi, once remarked, "Winning isn't everything; it's the only thing." That statement succinctly expresses the starting point for understanding all sports.

In natural evolutionary competition survival, to include reproduction, isn't everything; it's the only thing. An organism surviving and reproducing generation after generation in the struggle for existence is "winning" in evolutionary competition.

6.7   NATURAL ADAPTION

"How have all those exquisite adaptations of one part of the organization to another part, and to the conditions of life, and of one organic being to another being, been perfected?...we see beautiful adaptations everywhere and in every part of the organic world."

Charles Darwin, "On the Origin of Species"

In natural evolutionary competition, natural organisms must constantly change to survive and expand, or risk extinction. In natural evolutionary theory there are two main ways a natural organism can change in direct response to competitive or environmental conditions. The first, adaptation, utilizes the process of natural inheritance. The second, natural phenotypic plasticity, does not require the use of the process of natural inheritance. They are distinct.

6.7.1   THE PROCESS OF NATURAL ADAPTION

The term natural adaptation has more than one meaning in evolutionary theory. For simplicity's sake, we will use the term ***natural adaption*** to describe the process of natural adaptation in this book. Natural adaption is the mechanism by which natural organisms change in response to competitive and/or environmental changes. The change to natural organisms is produced through the process of natural inheritance. Therefore, natural adaption is accomplished through reproducing natural offspring over one or more generations.

An example is our ancestor's naturally adapting to walk upright. Our ancestors once walked on all four limbs like chimpanzees. However, at some point it became advantageous for our ancestors to walk upright. So, our distant ancestors began producing natural offspring that walked upright. This made our ancestors more evolutionary competitive. This is also an example of ***natural exaptation*** or the repurposing of an existing organ for a new, more beneficial purpose.

6.7.2 NATURAL ADAPTATION

A natural adaptation is a change in an organism's phenotype. This typically takes the form of structural or behavioral change. A natural adaptation occurs due to a specific stimulus in the environment. It typically removes or partially removes a natural check on a natural organism's survival or expansion. For example, a change in climate can cause a natural species to develop a new or lose an existing phenotypic trait such as fur. However, Darwin believed that the production of new natural adaptations was primarily driven by interaction and/or competition – both interspecies and intraspecies.

A natural adaptation is a change in a natural organism in response to competitive and/or environmental stimuli. An example of a natural adaptation is deer's antlers. Antlers are used by male deer to protect their mates and offspring. Antlers are also used to physically compete with other male deer for access to mating opportunities. This change provides natural evolutionary benefit that either improves a natural organism's chances of survival or its reproductive success.

Another term for this change is a ***natural adaptive trait***. Natural adaptive traits are beneficial as they enhance an organism's ability to compete in the struggle for existence. In natural evolutionary theory the changing of the human spine to enable walking upright is also called an adaptation or adaptive trait. This book will discuss two types of natural adaptations:

Natural Adaptation – Structural. Changes to an organism's system of organs is a structural adaptation. An organism's physical phenotype is comprised of its many structural adaptations. Examples are the human spine, deer antlers, and bird's beak.

Natural Adaptation – Behavioral. Change to an organism's instinctual behavior is a behavioral adaptation. Behavioral adaptations are the instincts organisms are born with. Examples are both the annual migration of birds and hibernation patterns of bears. As this concept is a bit more complex than structural adaptations, we will cover it in more detail below.

6.7.3   NATURAL INSTINCTS

"An action, which we ourselves require experience to enable us to perform, when performed by an animal, more especially by a very young one, without experience, and when performed by many individuals in the same way, without their knowing for what purpose it is performed, is usually said to be instinctive."

Charles Darwin, "On the Origin of Species"

Natural instincts are a species' inborn repetitive patterns of complex behavior. They are distinct from reflexes which produce variable responses to random environmental stimuli. Instead natural instinctive behaviors are hard coded and do not vary much across an entire natural species.

The simplest examples are fixed action patterns (FAP). FAPs have a defined sequence of actions customized to address a clear specific environmental stimulus. They are performed in response to specific environmental stimuli. In addition, there are other instinctual behaviors of more complexity and duration. Charles Darwin compared natural instincts to habits in his book *On the Origin of Species*:

"Frederick Cuvier and several of the older metaphysicians have compared instinct with habit. This comparison gives, I think, an accurate notion of the frame of mind under which an instinctive action is performed, but not necessarily of its origin."

Natural instincts are not learned but are naturally inherited behaviors. They are natural adaptive traits developed and passed on via the process of natural inheritance. An organ, such as the human brain, physically possesses hard coded instinctual behaviors in its wiring. Often natural instincts can be performed from the moment of birth. Other natural instincts develop as a natural species' offspring grow to maturity. An example is newly hatched sea turtles instinctively crawling to the ocean or the annual migration of birds to very specific destinations.

Some natural instincts can be improved with experience. In addition, other natural species can modify or restrain their patterns of behavior. This requires that a natural species has developed sufficient cognitive abilities to generate an internal stimulus (e.g., decisions, motivations, etc.). Human beings are capable of instinctual behavior pattern modification. Otherwise a natural species is locked into hard coded instinctual behavior even when detrimental to its immediate survival.

6.7.4   NATURAL CO-ADAPTATION

Natural co-adaptation is when two or more natural species develop natural adaptive traits beneficial only in the context of their interaction. We will address this concept in the section of Natural Perfection as it is best understood in context.

6.7.5   NATURAL PHENOTYPIC PLASTICITY

"The direct action of changed conditions leads to definite or indefinite results. In the latter case the organisation seems to become plastic, and we have much fluctuating variability."

Charles Darwin, "On the Origin of Species"

Natural phenotypic plasticity is an adaptive trait produced by the process of adaption. The benefit of this adaptive trait is 1) the ability to change the individual organism's phenotype and 2) to do so without the process of natural inheritance. In effect, the individual organism has the capacity to adjust some aspects of its natural phenotype (i.e., structural, behavioral, etc.) without a mutation of its DNA.

This flexibility removes a constraint imposed by the natural inheritance process – time. The organism can immediately initiate the phenotypic change in response to a direct environmental stimulus. This improves the chances of survival for both the individual organism and the species.

Examples of natural phenotypic plasticity is acclimatization. Human beings can reach high altitudes. However, they must allow sufficient time for their body to adjust to the change in altitude. Another example is mammals thickening and shedding of hair. Thickening in winter keeps mammals warmer and shedding in the summer keeps them cooler.

Wolves exhibit behavioral plasticity by adjusting their hunting/foraging strategies in response to the available food source. They eat prey of various kinds, carrion, fish, fruits, vegetables, and even other wolves. This flexibility of diet enables wolves to survive a diverse set of environmental conditions across Europe and North America.

6.8   NATURAL INHERITANCE

"We see in these facts some deep organic bond, throughout space and time...The bond is simply inheritance, that cause which alone, as far as we positively know, produces organisms quite like each other, or, as we see in the case of varieties, nearly alike."

Charles Darwin, "On the Origin of Species"

In its most basic sense the process of natural inheritance is about transmitting genetic information. One or more organisms (parents) pass on information to one or more replications or offspring. This transmitted information enables offspring to develop its parent's completed system of organs. This enables those offspring to develop a variation of its parent's form (i.e., structure, behaviors, etc.). An example of structure is the homo sapiens body. An example of instincts is a homo sapiens toddler's fear of strangers.

This information exchange during reproduction also enables natural species to change and adapt. In the struggle for existence a natural species' ability to change is a decisive factor. This makes the process of natural inheritance crucial in natural evolutionary competition. This process enables the efficient passing on of the pattern of natural adaptations to offspring.

6.9   NATURAL GENOMES

"We are survival machines – robot vehicles blindly programmed to preserve selfish molecules known as genes. This is a truth which still fills me with astonishment."

Richard Dawkins, "The Selfish Gene"

Natural genomes are at the core of natural reproduction which drives the process of natural evolution. In addition, natural inheritance and variation of genes is central to the process of natural evolution. A catholic friar, Gregor Mendel, discovered the patterns in how natural phenotypic traits are passed on to natural offspring from parents. He was the first to study genetics in a scientifically rigorous manner. The word genetics derives from the ancient Greek word for "genesis" meaning "origin".

Genetic or genomic information is organized in the below manner within homo sapiens. We will use Richard Dawkins' metaphor of how homo sapiens store information in books and libraries to simplify the explanation of this complex subject. This description is intended to only be a rough approximation.

DNA. This is the long double helix we are all familiar with. DNA is simply a single line of genomic information. The double helix is a set of two long lines of information. The information is instructions for producing an organism. Picture a book presented as just one long paragraph end-to-end. All the information necessary to create a human being is found in this long string of genomic information.

Nucleobases. These are the lowest level building blocks of DNA. Think of these as "letters". There are only four letters in the DNA alphabet: A, T, C, and G.

Nucleotides. These are the next level building blocks of DNA. Think of these as "words". There are only 64 combinations that nucleobases can be used to form nucleotides. That means that the DNA "dictionary" only has approximately 64 words to work with.

Nucleotide Sequences. The nucleotides are then organized into sequences. Think of these as like "sentences". The specific sequence of the nucleotides impacts how the DNA is read like a sentence in a paragraph.

Gene. A gene is a specific sequence of nucleotide sequences or "sentences" that are explicitly used to generate a natural phenotypic trait. Often multiple genes are combined to generate a genetic trait such as the homo sapiens eye. That is why a gene is the basic unit of inheritance. Each phenotype has an associated genotype that generated it. Think of genes as performing a similar function to "concepts" in homo sapiens thought.

Gene Sequence. The individual genes of a genotype are deliberately ordered in sequence to be read. If you change the gene sequence, you will begin to generate different adaptations and potentially a new organism. It is like an "outline" for an instructional manual. If you change the sequence or outline of concepts in an instructional manual, you change the meaning conveyed to the reader.

Genotype. The genotype of an organism is the combined genes and gene sequence of its DNA code. As said before, a genotype produces a specific phenotype. The genotype for homo sapiens produces the structure of the homo sapiens body and accompanying natural instincts. Think of the genotype as the "conceptual model" underlying an author's instructional manual. The author has ordered a specific set of concepts in a specific sequence to convey a specific meaning or main idea of his/her work. Genotypes work the same basic way.

Chromosome. Chromosomes are a single copy of the double helix DNA strands. Homo sapiens inherit 23 chromosomes from both their mother and father at conception. Chromosomes come in pairs making an "x" looking shape. Therefore, chromosomes contain 4 sets of DNA strands. Each set has the full instructions for producing an organism. Think of Chromosomes like "instruction manuals" that contain all the instructions for producing a natural organism.

Mutation. This is the alteration of the genome of a natural organism. Mutation is the source of genetic variation in organisms. Mutations occur due to copying errors or in response to specific environmental stimuli (e.g., competition, climate, etc.). Mutations can occur without producing new phenotypic traits in an organism's pattern. Mutations can also occur that alter one or more genes thereby producing a new adaptation in the natural organism's phenotype.

Nucleus. The nucleus of each cell in your body houses a complete set of chromosomes. Think of the nucleus of a cell as like a "library". All the "instruction manuals" are stored there for access in the future. When needed the information in your nucleus is used to produce new proteins. The new proteins are then used to produce new organs of a cell.

Cell. The cell is the smallest natural organism. Your body is comprised of over 30 trillion cells. Think of each cell as like a "room" and your body as the overall building. The patterns of the cells that your DNA instructed to create are the systems of your organs. The pattern of your system of organs is the pattern of your phenotype. Your first cell or zygote formed when a male sperm cell and a female egg cell came together.

In effect, homo sapiens start out as just a one "room building". The "bookcase" in that first room is a unique variation of the parents' "bookcase" – 23 books from each parent. Cells are created again and again until a fully formed homo sapiens is produced in nature.

6.10   SEXUAL SELECTION

"This leads me to say a few words on what I have called Sexual Selection. This form of selection depends, not on a struggle for existence in relation to other organic beings or to external conditions, but on a struggle between individuals of one sex, generally the males, for the possession of the other sex. The result is not death to the unsuccessful competitor, but few or no offspring."

Charles Darwin, "On the Origin of Species"

Sexual selection is a sub process of the process of natural selection. Many natural species have two distinct biological sexes – male and female. The two sexes participate in a mating process to reproduce natural offspring. The mating process involves two separate layers.

First, the males typically **compete** (intrasexual selection) with each other to secure access to the females. Second, the females typically **search** for the victorious males. Then females **select** from the victorious males (intersexual selection). The females are **selecting** the preferred male phenotype that led to the male's victory in intrasexual competition. In effect, the female is also **selecting** the underlying genotype that produced the visible phenotype.

Mating occurs once the female **selects** a victorious male. This is the process of passing genomic information from male to female. The female combines the male and female sets of genomic information. The female then produces natural offspring which are a unique variation of the parent's phenotypes.

6.11   NATURAL REPRODUCTION

"All vertebrate animals, all insects, and some other large groups of animals, pair for each birth...that is, two individuals regularly reunite for reproduction...But still there are many hermaphrodite animals which certainly do not habitually pair...it is a general law of nature that no organic being fertilises itself for a perpetuity of generations; but that a cross with another individual is occasionally – perhaps at long intervals of time – indispensable."

Charles Darwin, "On the Origin of Species"

Natural reproduction is the biological process by which organisms **produce** biological offspring. Natural reproduction occurs in two forms – sexual and asexual. Sexual reproduction includes the process of sexual selection. The female **searches** for a mate and then **selects** a specific male to breed with. The male then typically provides the female his genomic information (e.g., sperm cells). The female then **conceives** one or more new organisms.

The new organism is then typically **developed** in a womb or an egg. The offspring are then **produced** via the female giving birth, eggs hatching, etc. In asexual reproduction the organism simply replicates or clones itself to **produce** another organism with its own genomic information. The processes of sexual selection and sexual reproduction follow the same general pattern:

- The female **searches** for a suitable mate to impregnate her.
- The female **selects** a victorious male and his successful genotype.
- The male impregnates (transmits) genomic information to the female.
- The female **conceives** a new organism that blends both male and female genotypes.
- The female **develops** the new organism in a container such as a womb, eggs, etc.
- The female then **produces** the new organism via live birth, eggs hatching, etc.

This is the process by which most animal species on the planet are produced. Genomic information is transformed into a physical organism in nature via this process. Evolution has perfected this process over billions of years. It is the most efficient way on earth to produce new things.

6.12   NATURAL HYBRIDISM

"I have made so many experiments and collected to so many facts, showing on the one hand that an occasional cross with a distinct individual or variety increases the vigour and fertility of the offspring, and on the other hand that very close interbreeding lessens their vigour and fertility, that I cannot doubt the correctness of this conclusion."

Charles Darwin, "On the Origin of Species"

Hybridization is the process by which individuals from two distinct species mate. As a result of the process of natural inheritance the offspring produced will either be fertile or infertile. If the offspring are infertile then no further offspring will be produced. As the patterns of the two distinct forms of species were incompatible. However, if the offspring prove fertile then a new species, reproductively isolated from its parent species, will be created. This is a common phenomenon found in nature.

Hybridization is an extension of evolution's strategy of variation. The entirely new hybrid species combines two different genetic lines or patterns. This creates a new evolutionary trajectory for the new hybrid species. This offers new possibilities for adaptation that were previously unavailable in the parent species. Most adaptations facilitate the removal of constraints on a species' expansion in its ecosystem. This increases a species' probability of survival in the struggle for existence.

6.13   NATURAL VARIATION

"Nothing at first can appear more difficult to believe than that the more complex organs and instincts have been perfected...by the accumulation of innumerable slight variations, each good for the individual possessor."

Charles Darwin, "On the Origin of Species"

The process of reproduction involves the transmission of a copy of the parent(s) information to its offspring. It is like making a photocopy of a document today – often it is not an exact copy. In addition, in many species two organisms' mate to produce offspring. The reproductive process blends these two copies of information. Also, the information of an organism can change during its lifetime due to different causes.

As a result, each individual organism within a species has slight differences. Natural evolutionary theory terms these slight differences of organisms within a species a "variation". Variations express themselves in slightly modified organism form (i.e., structure, behavior, etc.). An example of variations is homo sapiens' different eye color and body height. Homo sapiens parents are often uncertain what eye color their offspring will express, or how tall their offspring will grow to be at maturity. It is to some degree random.

This randomness is a feature of natural evolution, not a bug. Often a random variation provides a new benefit to the species in the struggle for existence. This makes variation an important factor in the process of natural selection. Over time the accumulation of these new variations can lead to the development of a new distinct species.

6.14   NATURAL FITNESS

"This preservation of favorable individual differences and variations, and the destruction of those which are injurious, I have called Natural Selection, or the Survival of the Fittest."

Charles Darwin, "On the Origin of Species"

Every species is locked into a struggle for existence. And to "win" in that struggle a species must survive long enough to reproduce offspring that can also reproduce themselves. The more offspring created by each successive generation reduces the species' risk of extinction. Natural fitness is a relative term used to describe how effective a chosen genetic pattern is in the struggle for existence. A genetic pattern that produces more generational offspring that survive, relative to other genotypes, is considered to possess more fitness.

Fitness applies to a genetic pattern, not individuals. It refers to the expected average number of surviving offspring for a genetic pattern in each generation. This is the probability that the genetic pattern can reproduce an expected average of offspring in successive generations given stable environmental conditions.

This value of fitness is then used to assess the extinction risk for a particular genetic pattern. The lower the fitness level the higher the genetic pattern's extinction risk. The higher the fitness level the lower the genetic pattern's extinction risk. So, fitness is the term used to describe a genetic pattern's chances of "winning" in the struggle for existence. The species survives if just one genetic pattern survives.

6.15 NATURAL SELECTION

"As many more individuals of each species are born than can possibly survive; and as, consequently, there is a frequently recurring struggle for existence, it follows that any being, if it vary however slightly in any manner profitable to itself, under the complex and sometimes varying conditions of life, will have a better chance of surviving, and thus be naturally selected. From the strong principle of inheritance, any selected variety will tend to propagate its new and modified from."

Charles Darwin, "On the Origin of Species"

Charles Darwin gave a name to this process by which "winners" and "losers" are picked in evolutionary competition. He called this process natural selection. Basically, Charles Darwin theorized that species which "lose" in evolutionary competition are "deselected" for existence or go extinct. Those that "win" continues to access scarce resources and reproduce are "selected" for continued existence. Natural selection is the process by which species competitively self-sort themselves into either bucket – continued existence or extinction.

Evolution occurs by way of this self-sorting process. Over successive generations species create newer and hopefully better variations of themselves in the form of offspring. The improved and therefore more competitive offspring are then naturally selected for continued existence and reproduce themselves. This is evolution – the change in the inheritable characteristics of a species through successive generations. This is the evolution of a phenotypic pattern in a direction that maximizes a species' chances of survival given its phenotypic constraints.

Interestingly, variations of similar patterns are naturally selected again and again during evolutionary history. A great example of this is the pattern of the phenotypic form known as a crab. There are many variations of the phenotypic pattern of the crab on earth. But they are not all descended from the same ancestors. This means distinct genetic lines are evolving into roughly the same phenotypic pattern. This is consistent with Charles Darwin's observation in *On the Origin of Species* below:

"We can, in short, see why nature is prodigal in variety, though niggard in innovation. But why this should be a law of nature if each species has been independently created no man can explain...Many other facts are, as it seem to me, explicable on this theory...these facts cease to be strange, or might even have been anticipated."

Why is nature so "prodigal in variety, though niggard in innovation"? This is because the phenotypic pattern of the crab is advantageous in the survival of existence. This must mean that environmental and competitive patterns in evolution must be repeating as well. Otherwise the pattern of the crab wouldn't be a repeatedly advantageous form since the time of the dinosaurs.

We see this across the geological history of earth. Tetrapods or species with four legs (i.e., Ichthyostegas, triceratops, crocodiles, wholly mammoths, lions, etc.) have repeatedly evolved into existence for hundreds of millions of years. As the patterns of natural law and evolution continue to repeat, the same phenotypic patterns will tend to be naturally selected. As Charles Darwin stated in his book *On the Origin of Species*:

"It is generally acknowledged that all organic beings have been formed on two great laws – Unity of Type, and the Conditions of Existence...Hence, in fact, the law of Conditions of Existence is the higher law; as it includes, through the inheritance of variations and adaptations, that of Unity of Type."

6.16   COEVOLUTION

"...that the structure of every organic being is related, in the most essential yet often hidden manner, to that of all the other organic beings, with which it comes into competition for food or residence, or from which it has to escape, or on which it preys."

Charles Darwin, "On the Origin of Species"

Darwin repeatedly stated in his book that the relation between organisms is the single most important factor in evolution. Why? To "win" in the struggle for existence (i.e., exist, reproduce, expand) organisms must consistently obtain scarce natural resources. This resource constraint is a crucial driver of conflict between natural organisms. Individual organisms compete for food, territory, and to reproduce. Some natural organisms (predators) hunt other natural organisms (prey) as their food resource.

This organism-to-organism (e.g., predator-prey, etc.) competition leads to the process of coevolution. Coevolution is the reciprocal evolutionary process. It occurs when two or more species have sustained interaction within a natural ecosystem. The dynamic of interaction creates a persistent selective pressure on both species. This selective pressure then directly affects both species' evolutionary trajectory.

Coevolution often occurs as ever escalating 'arms race' between competing organisms. An example of this is the predator-prey relationship between bats and moths. Bats developed the adaptive trait of echolocation to detect moths in the dark. Moths in turn developed an adaptive trait of ears with a greater hearing range. This serves the moths as an early warning system to hear bat echoes from far away. This process of coevolution often leads to species becoming very specialized. This specialization can then lead to the development of new species over geological time.

6.17   NATURAL CO-ADAPTATION

"Thus I can understand how a flower and a bee might slowly become, either simultaneously or one after the other, modified and adapted to each other in the most perfect manner, by the continued preservation of

all the individuals which presented slight deviations of structure mutually favorable to each other."

Charles Darwin, "On the Origin of Species"

Co-adaptation is when two or more species develop new interdependent adaptive traits. The adaptive traits are beneficial only in the context of their interaction. In effect, the species involved adapt their individual patterns (i.e., structure, behavior, etc.) to integrate with each other. This creates evolutionary value otherwise not available to either species. This increases both species' chances of being naturally selected in the struggle for existence.

A simple example of this is the interactions between bees and flowers. Both species have developed adaptive traits that benefit the other in the struggle for existence. Flowers have developed color and form to make it easy for bees to spot them. Bees farm flower pollen to make honey which feeds their hive. Bees have developed hair on their body to become more efficient at collecting pollen.

Flowers want as much pollen to be collected and spread as this is how they reproduce. The bees drop pollen flying back to the hive fertilizing other flowers with the pollen. It is a win-win scenario for both species.

6.18   NATURAL PERFECTION

"Natural selection tends only to make each organic being as perfect as, or slightly more perfect than, the other inhabitants of the same country with which it comes into competition. And we see that this is the standard of perfection attained under nature...Natural selection will not produce absolute perfection..."

Charles Darwin, "On the Origin of Species"

Natural perfection is a term used to describe the existing state of a species' phenotypic pattern. A species phenotypic pattern is, as Richard Dawkins says, only a "survival machine". It is only a means to an end, not an end in itself. Therefore, the perfection of a phenotypic pattern can only be understood relative to its specific end. That specific end is maximizing the chances of a species surviving and reproducing in its specific natural ecosystem.

Natural ecosystems are made up of many interdependent natural patterns (i.e., natural laws, competitive, weather, solar, other species, etc.) as well. A species' phenotypic pattern is said to be perfected relative to the patterns of its natural ecosystem. So, there can be no absolute pattern of perfection in natural evolution – only a pattern of perfection relative to other existing patterns. But what does it mean for a species to be perfected?

A species has reached a state of natural perfection when it has no evolutionary incentive to adapt any further. This means that any further phenotypic adaptations would only confer a cost without any immediate benefit. In addition, the new adaptation will likely require more nutriments to support its existence. If a species kept evolving, it would in effect be evolutionarily wasteful. Over geological time even the slightest waste could cause a species' eventual extinction.

So, a species will stop evolving along its current evolutionary trajectory unless there is a "pull" of an immediate benefit present in the ecosystem. This can be in the form of a new resource that can be exploited (offense) or a response to a competitor's new adaptation (defense).

Perfection is an evolutionary value break-even point. It occurs when a species has perfected relative to the constraints (e.g., resources, competition, environmental, etc.) present in its ecosystem. All species evolve in this manner. It is due to the dynamic and uncertain nature of evolutionary conditions. In this context it makes little sense for a species to make any long-term investments. This makes evolution the ultimate day trader.

6.19   NATURAL SPECIATION

"Consequently, in the course of many thousand generations, the most distinct varieties of any one species of grass would have the best chance of succeeding and of increasing in numbers, and thus supplanting the less distinct varieties; and varieties, when rendered very distinct from each other, take the rank of species."

Charles Darwin, "On the Origin of Species"

Speciation is the process by which a variation of a species becomes a distinct species of its own. This occurs when a group of similar organisms (i.e., a variation) within a specific species begin to evolve along an evolutionary trajectory distinct from the parent species. Over time the variation's adaptive traits become distinctly different than the parent species. This typically enables the variation to attain an increased level of fitness relative to its specific ecosystem. Species diversify wherever possible to reduce extinction risk. It is this process of diversification that eventually causes speciation.

Natural ecosystems present similar conditions and dynamics to existing and new species. In addition, all natural ecosystems are governed by natural laws of the universe such as gravity. Therefore, the evolution of each species follows the same general pattern - speciate, vary, expand, stasis, contract, go extinct or speciate again. It is estimated that over five billion species have existed on the planet earth. But 99% of them have become extinct by generally the same pattern.

6.20   NATURAL EQUILIBRIUM

"Battle within battle must be continually recurring with varying success; and yet in the long-run the forces are so nicely balanced, that the face of nature remains for long periods of time uniform, though assuredly the merest trifle would give the victory to one organic being over another... the incessant action and reaction of various forces, which, as throughout nature, are always tending towards an equilibrium..."

Charles Darwin, "On the Origin of Species"

Equilibrium in natural ecosystems occurs when a persistent set of competitive and environmental patterns emerge. This does not mean the ecosystem loses its dynamic nature – but that the changes happen within the same persistent patterns that recur over time. This creates relative stability in the ecosystem in terms of competitive and environmental dynamics. This relative stability exists for a protracted period. When this relative equilibrium is reached predictable patterns will recur over time in response to recurring events such as seasonal change, droughts, floods, migrations, resource fluctuations, etc. All species will become perfected within this overarching set of persistent patterns. This relative equilibrium will persist in a natural ecosystem until a disruption occurs.

6.21 NATURAL DISRUPTION

"But we have better evidence on this subject than mere theoretical calculations, namely, the numerous recorded cases of the astonishingly rapid increase of various animals in a state of nature, when circumstances have been favorable to them during two or three following seasons."

Charles Darwin, "On the Origin of Species"

Natural disruptions are events that permanently alter one or more persistent patterns of a natural ecosystem. The source of the disruptive event can be either internal or external. Internal disruptive events typically occur when an existing constraint is either removed or becomes more restrictive. An external disruptive event occurs when something new, such as an invasive species, impacts the patterns of the natural ecosystem.

Examples are a natural species within a natural ecosystem develops a new natural adaptation. This new natural adaptation enables the species to enhance access to an existing resource (i.e., territory, food, etc.). A period of turbulence will then occur in the natural ecosystem as an altered dynamic of competition arises. The natural ecosystem will then settle into a new form of relative natural equilibrium. This relative natural equilibrium will hold until the next disruptive event.

6.22 NATURAL PUNCTUATED EQUILIBRIUM

"Evolution is a theory of organic change, but it does not imply, as many people assume, that ceaseless flux is the irreducible state of nature… Change is more often a rapid transition between stable states than a continuous transformation at slow and steady rates."

Stephen Jay Gould, "The Panda's Thumb"

In 1972, paleontologists Stephen Jay Gould and Niles Eldredge published a scientific paper called *Punctuated Equilibria*. In this paper Gould and Eldredge presented a new theory in evolutionary science. Charles Darwin's theory of evolution presumed that the process of evolution generally preceded gradually over time.

Gould and Eldredge theorized that the evolution of species would remain relatively unchanged for long periods called stasis. This period of stasis would then be followed by a period of extreme evolutionary volatility. This would cause rapid populational growth of some species while simultaneously causing the extinction of other species.

Once the period of extreme volatility ended another long period of relative statis would commence. Then Darwin's gradual evolutionary process would start again. Gould and Eldredge theorized that this pattern had repeated throughout earth's history. They called it punctuated equilibrium.

6.23   "THE ONLY GAME IN TOWN"

"We meet with this admission in the writings of almost every experienced naturalist; or as Milne Edwards has well expressed it, Nature is prodigal in variety, but niggard in innovation. Why, on the theory of Creation, should there be so much variety and so little real novelty?"

Charles Darwin, "On the Origin of Species"

Charles Darwin asked the above question several times in his book. At the time he wrote *On the Origin of Species,* he did not yet have knowledge of natural genetics. But today we know that species' evolutionary possibilities are generally constrained by their genotype. This then places a constraint on the possibilities for a species' phenotypic pattern. But that is only a partial answer. The full answer lies in another theme Charles highlighted which is captured in the quote below:

"It is generally acknowledged that all organic beings have been formed on two great laws – Unity of Type, and the Conditions of Existence...Hence, in fact, the law of Conditions of Existence is the higher law; as it includes, through the inheritance of variations and adaptations, that of Unity of Type."

Our planet's evolutionary history has been shaped by patterns. The pattern of genotype is only one pattern that shapes the evolution of all species. There is one pattern of natural laws (i.e., gravity, motion, etc.), one pattern of environmental conditions (i.e., weather, geology, etc.), and one pattern of competition between organisms (i.e., coevolution, co-adaptation, perfection, etc.).

These individual patterns have varied in their specific form over billions of years. But these patterns have existed in some form throughout earth's history. These patterns combine to form one single, integrated pattern – a pattern that cannot be physically seen. But a pattern that has caused the existence of everything that can be seen.

And this one single pattern has never varied in its form. It has shaped the evolutionary reality of the Jurassic Period, the Ice Age, and modern-day Africa. Each of these specific ecosystem forms are just variations of each other – all created by the one underlying pattern. That is why the same ecosystem and phenotypic patterns have repeated throughout earth's history. As the French writer Jean-Baptiste Alphonse Karr once wrote:

"The more things change, the more they stay the same."

The statement is seemingly paradoxical until you realize that all evolutionary history is a repetition of the same patterns in varied forms. Even when an asteroid struck the planet the one single pattern just shrugged it off and repeated another variation of itself. This means everything you see today is a variation of past evolutionary realities. That includes all creations of human civilization. It must be a variation. This one single pattern has repeated for billions of years. This means that on earth the one single pattern is:

The only game in town

7

CHAPTER 7: ARTIFICIAL SELECTION

"Slow though the process of selection may be, if feeble man can do much by artificial selection, I can see no limit to the amount of change, to the beauty and complexity of the co-adaptations between all organic beings... their physical conditions of life...effected in the long course of time through nature's power of selection, that is by survival of the fittest."

Charles Darwin, "On the Origin of Species"

7.1   SELECTIVE BREEDING

"In Saxony the importance of the principle of selection in regard to merino sheep is so fully recognized, that men follow it as a trade: the sheep are placed on a table and are studied, like a picture by a connoisseur; this is done three times at interval of months, and the sheep are each time marked and classed, so that the very best may ultimately be selected for breeding."

Charles Darwin, "On the Origin of Species"

In Charles Darwin's theory of natural evolution artificial selection is a substitute mechanism for natural selection. Instead of natural selection self-sorting species for existence or extinction, humans interfere intentionally to make the selection.

This is also known as the process of selective breeding in which humans selectively breed animals and plants. We do so to produce phenotypic traits beneficial to us, not necessarily beneficial to the organisms. This form of artificial selection has been practiced by homo sapiens for thousands of years.

A key characteristic of artificial selection is that it accelerates the process of adaption. This is due to homo sapiens leveraging our imaginative capacity. We deliberately select organisms we think will accelerate the evolution of a species down a selected trajectory – one that will produce the phenotypic traits we are intending to create.

Dog breeding is a great example of artificial selection. Our ancestors bred wolves using the process of artificial selection. Over thousands of years we gradually domesticated some wolves. This drastically changed their behavior and appearance into a new species – dogs. To this day dogs are still man's "best friend".

7.2   AN ANOMALY TO THE PATTERN

"One of these things is not like the others; One of these things just doesn't belong; Can you tell me which thing is not like the others; by the time I finish this song?"

Joan Ganz Cooney, "Sesame Street Lyrics"

The concept of artificial selection is an anomaly to the pattern of Charles Darwin's theory of natural evolution. At no other point does he discuss any meaningful involvement of homo sapiens in the natural evolutionary process. Generally he has homo sapiens viewing nature from the outside looking in. In addition, Charles often requests the reader to abandon our myopic notions of time and space. Otherwise comprehension of how natural evolution functions is difficult to achieve.

The answer to the anomaly lies with the natural species homo sapiens itself. Our natural species is an anomaly in natural evolutionary history as well. What makes us anomalous? It is our capacity for conscious thought and imagination. As the French mathematician and philosopher Rene Descartes once said:

"I think, therefore I am."

But in this case, we think, therefore we are an evolutionary anomaly. But then what is artificial selection really? It seems to be a combination of conscious thought followed by a physical action that produces an intended "thing". To find the answer to our question we must follow the intellectual breadcrumbs left behind by both Charles and Erasmus Darwin.

7.3 THE PATTERN OF FAMILY DARWIN

7.3.1 CHARLES DARWIN

"Linnaeus and Cuvier have been my two gods, though in very different ways, but they were mere schoolboys to old Aristotle."

Charles Darwin, "The Life & Letters of Charles Darwin"

Charles Darwin revered the ancient Greek philosopher Aristotle. He was the father of modern biology and the first known person to systematically document patterns in nature. Aristotle published many books on biology as well as on logic and metaphysics. His work included identifying and classifying over 500 distinct species of animals.

Aristotle's writings were somewhat close to articulating the pattern of Charles Darwin's theory. But Aristotle got several crucial concepts in the pattern wrong. This is because he had to work mostly with his own imagination and observational skills. In Aristotle's time empirical scientific information was very limited. But Charles Darwin was not the only Darwin with a keen interest in ancient Greek philosophy. His grandfather, Erasmus Darwin, was a well-known philosopher.

7.3.2   ERASMUS DARWIN

"So erst the Sage [Pythagoras] with scientific truth...In Grecian temples taught the attentive youth; With ceaseless change how restless atoms pass [Democritus]...From life to life, a transmigrating mass...How the same organs, which to-day compose...The Poisonous henbane, or the fragrant rose...May with to-morrow's sun new forms compile...Frown in the Hero [Socrates], in the Beauty smile [Plato]...When drew the breath enlighten' d Sage [Aristotle] the moral plan...That man should ever be the friend of man...Should eye with tenderness all living forms...His brother-emmets, and his sister-worms."

Erasmus Darwin

Erasmus Darwin was a polymath with philosophy being one of his many academic interests. The excerpt above is from his poem "To the Stars". In the poem it appears Erasmus references the ancient Greek philosophers Pythagoras, Democritus, Socrates, Plato, and Aristotle. I have added brackets in the text to highlight where in the poem. Erasmus then ended the poem with advice to man to be kind to other species because we all descend from common ancestors.

Erasmus published many books and poems on a variety of subjects. Many of his published texts expressed the concepts of ancient Greek philosophy in some form. So, it appears that both Erasmus and Charles venerated the ancient Greek philosophers. They both drew insights and inspiration from the ideas of arguably the greatest minds in history. Ancient Greek ideas heavily influenced their respective works. But what exactly is an idea?

7.4   THE ANCIENT GREEK IDEA

"And we also assert that there is a fair itself, a good itself, and so on for all things that we set down as many. Now, again, we refer to them as one idea of each as though the idea were one; and we address it as that which really is...That's so...And, moreover, we say that the former are seen, but not intellected, while the ideas are intellected but not seen."

Plato, "The Republic"

It turns out that the term idea is derived from ancient Greek word *"ἰδέα"* meaning "form, pattern". The word *"ἰδέα"* is then derived from the root ancient Greek word *"ἰδεῖν"* meaning "to see." The word "form" can also be translated as the word **"φύσις"** or "nature" in ancient Greek.

If you combine the terms, you get the phrase "to see a pattern in nature". Isn't that what Erasmus and Charles were doing – seeing patterns in nature? So, that fits our overall pattern – natural evolution is the driving force at the center of nature.

Ideas are conceived through the process of conscious thought. Artificial selection is a combination of conscious thought and intentional action that produces a specific "thing". The ancient Greek philosophers caused a revolution in homo sapiens thought almost 2,500 years ago – and both Erasmus and Charles drew insights and inspiration from their ideas. So, it seems we must turn to the ancient Greek philosophers to discover the true meaning of the term artificial selection.

8

CHAPTER 8: REFRAMING THE PROBLEM

"To return from nature to φύσις [nature] is to venture to suspend this history so as to retrace the figure that oriented philosophy in its Greek beginning. It is to venture the attempt to write again περὶ φύσεως [on nature], to span the distance in such a way that it might become possible from this distance nonetheless to reinscribe such discourse."

John Sallis, "The Figure of Nature: On Greek Origins"

8.1   THE ANCIENT GREEK REALITY

"Everything comes back somehow or other to nature. All things return to it along some way. For every thing is, if not nature itself, nonetheless a thing of nature, a natural thing; nothing is completely apart from nature, not even the gods. Indeed, it is primarily in and through nature that the gods make their presence known, to such an extent that their very presence is inseparable from the manifestations of nature."

John Sallis, "The Figure of Nature: On Greek Origins"

The ancient Greek reality was very different than that of our modern world. Ancient Greece is today famous for the individual polis or city state. Each city state was organized around a single urban center like national capitals today. We can identify with that societal reality as each year more people live in urban areas.

However, the everyday reality for most Greeks was far different. It was closer to that of the traditional Maasai society in Kenya. The Maasi are pastoralists and cattle herders. They are also fierce warriors. The Greeks were also highly skilled farmers. Greece is world famous for its tradition of olive farming started 5,000 years ago. Ancient Greece was much closer to that of the traditional Maasi society then modern homo sapiens society.

In addition, the Greek system of faith was inseparable from the natural world. The Greek gods were thought to literally be present in nature itself. Examples are Zeus the god of sky and thunder, Poseidon the god of the sea, storms, earthquakes and horses, Apollo the god of the Sun and light, Demeter the goddess of the harvest, and Artemis the goddess of the wilderness, animals, and the Moon.

To the Greeks nature was literally divine – the truth of the gods was concealed within nature itself. As a result, the ancient Greeks drew no intellectual boundary between nature and the city-state. Such a thought would have been incomprehensible to most ancient Greeks. This ancient Greek reality is the context within which what we term ancient Greek philosophy must be understood. It predated our concept of philosophy itself.

8.2   ANCIENT GREEK PHILOSOPHY

"In this μῦϑος [myth] the figure of nature was already drawn before philosophy came onto the scene and set about interrogating nature as such...Even though the name Artemis goes largely unmentioned by the early Greek thinkers, the disclosure of nature sustained by her μῦϑος [myth] remained directive for Greek thought from its beginning on."

John Sallis, "The Figure of Nature: On Greek Origins"

The term philosophy was derived from the combination of two ancient Greek words. The first "φίλος" or "philos" meaning "love" and the second "σοφία" or Sophia meaning "wisdom". Together they form the term philosophy meaning "love of wisdom". The word philosophy did not enter modern language (i.e., English, French, etc.) until approximately 1175 A.D. It was based on the Latin term "philosophia" which had a narrower academic scope and meaning than that of the Greeks.

The scope of ancient Greek Philosophy was roughly equivalent to every department in a modern-day university. Philosophy was the systematic enquiry into subjects such as astronomy, epistemology, mathematics, politics, ethics, ontology, logic, metaphysics, biology, rhetoric, technology, etc. Except philosophy drew no sharp distinctions between the various academic fields.

The goal of philosophy was to reveal the truths of nature and thereby the secrets of the gods. To separate the academic fields would be equivalent to separating the gods. Such a suggestion would have been foreign to the ancient Greek mind – they lived in a world driven by the everyday interactions of the gods expressed in nature. And even as philosophy evolved away from mythology toward rationality, nature would forever remain the context in which that rationality would be exercised.

8.3   THE ANCIENT GREEK PHILOSOPHERS

"According to ancient testimony, it was Pythagoras who first called himself φιλόσοφος [philosopher]…Only later…did the thinkers of ancient Miletus come to be designated philosophers…no doubt opened a way of access to their thinking, especially to the degree that the primary sense of philosophy continued, as with the Milesians, to be determined by an orientation to φύσις [nature]."

John Sallis, "The Figure of Nature: On Greek Origins"

Homo sapiens that strove to reveal the truths of nature called themselves philosophers or "the lovers of wisdom". In so doing they drew no sharp distinction between nature and human activity. It would be equivalent to separating themselves from the gods and thus this would be considered sacrilegious. Quite the opposite, philosophers were actively attempting to commune with the gods through nature. That is why Aristotle, a renowned Greek philosopher, repeatedly describes perfected human activity as "god-like".

The thinking of philosophers evolved over time. The earliest philosophers were generally focused on revealing the secrets of natural phenomena. After Socrates, considered the greatest Greek philosopher, they began enquiring into unique human subjects such as thought and politics.

However, nature, and therefore the gods, were always the context within which this philosophic evolution took place. This is reflected in both the writings of Greek philosophers and the language they used to write them. The philosopher's intellectual center of gravity was always the natural world.

8.4 "WRITINGS ON NATURE"

"For those who wrote in proximity to the beginning of philosophy, the primary focus and animating theme of their thought was, with only few exceptions, φύσις [nature]. Even later, when the venture is launched to set thinking apart from the beginning – as in Socrates' second sailing – φύσις [nature] remains the reference point from which whatever might be projected beyond would be determined."

John Sallis, "The Figure of Nature: On Greek Origins"

Most early philosophical writings were centered on nature. By nature they meant the entire world to include the gods. In fact, many of the writings were explicitly titled "On Nature". Some other writings were later given a title associated with nature by librarians at places such as Alexandria in Egypt. The librarians did so as it was apparent to them the subject of the writing was nature. The quest to reveal the secrets in nature was both a starting reference point and contextual constraint for philosophic thought's evolutionary trajectory.

8.5 ANCIENT GREEK LANGUAGE

"To return from nature to φύσις [physis or nature] is not merely to substitute for a modern word or concept ancient equivalent. Rather, it is to reverse a history of translation that, beginning with the Latin rendering of φύσις as natura [nature], has distanced what is said in the translation from what was once said in the word φύσις."

John Sallis, "The Figure of Nature: On Greek Origins"

The very language used by the ancient Greek philosophers to describe philosophical thoughts was also centered on nature. The word "truth" in ancient Greek is "ἀλήθεια" or "Alethia" meaning revealing, unconcealedness, cr the state of not being hidden. It also means "factuality or reality". The antonym is "lethe" or concealment.

For the ancient Greek philosophers discovering the truth or facts of reality was the act of revealing the truths of nature and therefore of the gods. The word "ἰδέα" idea meaning "form, pattern" and its root of "ἰδεῖν" idein meaning "to see" is the clue. Another meaning of the word "form" in ancient Greek is "nature".

This gives the new meaning to the word idea of "to see a pattern in nature." If you pair this with the meaning of alethia you get "to see a pattern in nature that reveals the truth of the gods." This was the goal of ancient Greek philosophy itself. This makes the above translation not only plausible, but likely correct given the context. We will assume it is going forward.

8.6   LOST IN TRANSLATION

"But if thought corrupts language, language can also corrupt thought."

George Orwell "1984"

Now, take the single mistranslation of the word idea and multiply that millions of times. That is what has happened over the last two thousand-five hundred years. The meaning of the words has in many cases, been passed thru multiple languages, empires, religions, geographic regions, academic institutions, individual scribes, etc. The way that many of these classic Greek texts have passed down to us is the exact opposite of how historians would have wished they had.

In effect, through the centuries ancient historians have had to be like Sherlock Holmes. In some cases, they have had to solve extremely difficult puzzles to obtain the simplest pieces of information. We should collectively applaud the academic community for the breakthroughs made thus far. It is quite literally an impossible task given the myriad of potential points for quality error during this process. This is why over two thousand years later we have still yet to decode parts of ancient Greek text accurately. This is despite herculean efforts by institutions and individuals throughout history.

This has all led to the corruption of the information passed through the successive rounds of direct translation. It also has had the subtle effect of corrupting our thoughts. For that corrupted information has been a constraint on modern philosophic thinking – a constraint that homo sapiens have been trying to work around for millennia. It is this constraint we must remove to decode the truths hidden in ancient Greek philosophical texts. This is a constraint not in language, but in the mind.

8.7   REFRAMING THE PROBLEM

"Alexander looked at an island and saw, instead, land. He reconceived his problem so completely as to render pregnable what was impregnable. He reframed his problem from one of naval to one of earthly proportions. While such minds are rare, emulating them need not be."

Lance B. Burk, The Wisdom of Alexander the Great

If direct translation has not fully solved the problem, then another way must be found. But we must now abandon the point of view previously described above. For as the famous theoretical physicist Albert Einstein is thought to have once said:

"We cannot solve our problems with the same thinking we used when we created them."

Luckily, we can turn to the ancient Greeks to help us to decode the ancient Greeks. Arguably the philosopher Aristotle's best student was not a philosopher but a king. After the death of Plato in 343 B.C. Aristotle traveled to Pella, the capital of Macedonia. The Macedonian king had personally requested Aristotle to tutor his young son. The father was Philip, and his son became who we know today as – Alexander the Great.

That's right – Socrates taught Plato, Plato taught Aristotle, and Aristotle taught Alexander.

In his book *The Wisdom of Alexander the Great,* Dr. Lance Burke describes the Aristotelian thinking Alexander used to conquer the Persian Empire and beyond. The first quote in this section describes Alexander's conquest of the island city of Tyre. Alexander didn't have a navy so he couldn't conquer an island.

So Alexander conquered the island by making it **_NOT_** an island. The Macedonian army built a mole or land bridge out to the island – then stormed the city. Today, over two thousand years later, the city of Tyre is still part of the mainland. The mole was used to build a stone causeway that stands to this day. In his book, Dr. Burke describes the leadership process Alexander followed at Tyre:

"When confronted with a seemingly unsolvable problem, you can reframe the problem, solve that new problem, and eliminate the original problem. I call this problem displacement."

This is how we will solve the problem of language translation that has plagued ancient historians for millennia. We won't solve that problem but displace it instead. We will select a new problem to solve and thereby make the original problem obsolete. But where can we turn to a model for solving such an extremely difficult puzzle? For that we turn back to the father of artificial intelligence – Alan Turing.

8.8   AN ENCODED MESSAGE

"Codes are a puzzle. A game, just like any other game."

Alan Turing, The Father of Artificial Intelligence

During World War 2 Alan Turing joined the British Military Intelligence at Bletchley Park in Buckinghamshire, England. He was recruited to contribute to the attempt to break German messaging codes. The German High Command had developed a cipher device named Enigma. The Enigma Machine was used to protect the most top-secret messages.

The movie *Imitation Game* is about the spirit of Alan Turing's contribution to the successful cracking of the Enigma code. Alan was a major contributor to the team that cracked the Enigma code. That gave a decisive competitive edge to the Allies. This likely shortened the war and saved untold lives. While many of the scenes of the movie may not be factually correct, the steps the characters take to crack the code can still be useful.

In the movie the Germans make a simple mistake that enables the Allies to crack the code. While they varied the encoding key each day, they did not vary the text of many of the transmitted messages. Specifically, they did not vary the format of their weather reports. They often repeated the same words and phrases. The weather reports all ended with the phrase: "Hail Hitler!". This lack of pattern variation was an inherent weakness in the Enigma code process.

The allies built a new machine, later called the Turing Machine, which enabled them to rapidly decrypt messages. Together with the weakness of pattern variation in German messages it enabled the allies to daily read their messages from 1943 until 1945. We are also attempting to decode a message – except ours is a static message. All we need to get started is a cipher that enables us to decode part of the message. The family Darwin have left us the clue we needed.

8.9   THE ULTIMATE CIPHER

"We can, in short, see why nature is prodigal in variety, though niggard in innovation. But why this should be a law of nature if each species has been independently created no man can explain...Many other facts are, as it seem to me, explicable on this theory...these facts cease to be strange, or might even have been anticipated."

Charles Darwin, "On the Origin of Species"

The natural world has remained relatively unchanged since the time of Classical Greece or the Victorian Era. As Charles Darwin states above "nature is prodigal in variety and niggard in innovation". Therefore, both the Greek philosophers and the Darwins were looking at the same general reality. Based on the writings of both the Darwins, it seems they leveraged concepts from the ancient Greeks to crack the code of the theory of natural evolution.

If this hypothesis is true, then the reverse must also be true. We should be able to use the theory of natural evolution as originally expressed in *On the Origin of Species* to decode the texts of ancient Greek philosophy. This makes the conceptual pattern embedded in the Darwins' theory of natural evolution our cipher.

8.10   DECODING THE MESSAGE

"In fact the a priori reasoning is so entirely satisfactory to me that if the facts won't fit in, why so much the worse for the facts is my feeling."

Erasmus Darwin, English Philosopher

This cipher alone will likely not be sufficient to decode the message. Even if we follow Erasmus Darwin's line of thinking below, we might be short some meaning to validate our hypothesis. If so, it will require that we leverage homo sapiens' most powerful adaptive trait. As Albert Einstein once described this adaptive trait:

"Imagination is more important than knowledge. Knowledge is limited. Imagination encircles the world."

Once the explicit facts come up short, we will use our imagination. We will use our imagination within the context of the ancient Greek philosopher's perspective. This is the intellectual quest "to see a pattern in nature that reveals the truth of the gods." We already know the patten in nature – evolution.

So, in the next chapter we will combine the conceptual pattern of the theory of natural evolution with our imagination. We will use these assets to validate our *a priori reasoning* or reasoning developed independently from any direct experience. This type of *a priori* pattern matching reasoning is at the core of Plato's writings. So, we will select his dialectic *Meno* as our test case in the next chapter.

9

CHAPTER 9: DECODING THE MENO

"To return from nature to φύσις [nature] is to venture to suspend this history so as to retrace the figure that oriented philosophy in its Greek beginning. It is to venture the attempt to write again περὶ φύσεως [on nature], to span the distance in such a way that it might become possible from this distance nonetheless to reinscribe such discourse."

John Sallis, "The Figure of Nature: On Greek Origins"

9.1 INTRODUCTION TO MENO

"Thus while the first part of Meno...looks very like a dialogue of search, tackling the question 'What is excellence?', the bulk of the dialogue raises fruitful questions which are designed to overcome difficulties raised by the search."

Robin Waterfield, "Plato: Meno and Other Dialogues"

The *Meno* is a Socratic dialogue by Plato. The dialogue takes the form of a dialectic. The dialectical method is a dialogue between people who hold different points of view on a subject. The goal of the dialectic is to arrive at the truth by reasoned argument. It differs from a debate as dialectics exclude elements such as emotional appeal and rhetorical argument. The focus is on the purity of the logic itself to persuade the reader of its truth.

The truth the dialectic *Meno* attempts to search for is human excellence – the ultimate virtue. The characters in the dialogue engage in the Socratic Method to search for the meaning of human excellence. Part of this process is defining exactly what each concept really means. The dialogue then ends with the characters still not knowing what exactly virtue is, but only how a human acquires it – from the gods.

In truth, however, Plato has given the reader the answer – the greatest human virtue is imagination. The answer is subtly weaved into the dialogue itself. Plato has subliminally communicated the pattern of excellence.

9.2   SOCRATIC METHOD

"In short, then, the truth which Socrates searches for by means of the elenchus [Socratic method] is the kind of truth which accompanies consistency. If a consistent set of beliefs, which incorporates notions... which are reasonably held to be true, survives repeated elenchi [Socratic questionings], it has a better chance of being true than an inconsistent set."

Robin Waterfield, "Plato: Meno and Other Dialogues"

Socrates is famous for assuming he knew nothing. According to Plato the Oracle at Delphi once stated that "Socrates is the wisest person in Athens." But Socrates believed he knew nothing. Logically this meant that nobody in Athens knew anything. Socrates was the wisest because he admitted that he knew nothing. Everyone else in Athens had opinions that they thought were true.

The Socratic Method is based on this understanding of humanity's instinctual pattern matching ability. This is why it is more effective in a group setting. The group pools its pattern matching ability and accelerates the process of drilling down to a concept's pure pattern. But the key element in the Socratic Method is the humility that the participants express. They assume that they may really know nothing about the subject of enquiry. If all participants in the enquiry exhibit this humility, then the pattern-matching process is accelerated dramatically.

9.3 REDEFINING EXCELLENCE

"How have all those exquisite adaptations of one part of the organization to another part, and to the conditions of life, and of one organic being to another being, been perfected?...we see beautiful adaptations everywhere and in every part of the organic world."

Charles Darwin, "On the Origin of Species"

The ancient Greek word **"ἀρετή"** means virtue or excellence. It means fully maximizing the potential of a "thing" to perform its inherent function – man, chimney, horse, a city-state, etc. Each "thing" has its own unique form of excellence. The focus of the *Meno* is discovering the unique form of excellence for a specific "thing" – homo sapiens.

Plato believed the greatest form of homo sapiens' excellence is the exercise of imagination. Imagination is the single natural adaptation that enables homo sapiens to adapt to any circumstance or situation. It allows for homo sapiens to generate knowledge of the patterns in nature. All other human capabilities derive from the natural adaptation of imagination and its pattern-matching capacity. This is the excellence the characters search for in *Meno* – imagination.

9.4 MENO

9.4.1 WHAT IS HUMAN EXCELLENCE?

"<u>MENO:</u> I wonder whether you can tell me, Socrates, whether excellence is teachable or, if not teachable, at least a product of habituation. Or perhaps it isn't a kind of thing one can practise or learn, but is a natural endowment. If not, how do people become good [excellent]?"

Plato, "Meno"
Robin Waterfield, "Plato: Meno and Other Dialogues"

In *Meno* the author, Plato, seeks to discover how homo sapiens become excellent. Three possibilities are proposed by Plato: 1) teaching, 2) habituation, 3) natural inheritance. As previously discussed, Meno is the perfect foil for this enquiry. Meno is a man who possesses no excellence, so he is perfect person to help discover what excellence is not.

Plato subtly tells the audience that neither Meno nor Gorgias know what excellence is when he says they share the same "opinion". Since knowing what something is not necessary in defining something, Meno and Gorgias are useful instruments in the search for excellence. In addition, Plato foreshadows the conclusion of the dialectic. He has Socrates state that he must be "high in the gods' favour" to know the answer.

Meno. Can you tell me, Socrates, whether virtue is acquired by teaching or by practice; or if neither by teaching nor practice, then whether it comes to man by nature, or in what other way?

Socrates. O Meno, there was a time when the Thessalians were famous among the other Hellenes only for their riches and their riding; but now, if I am not mistaken, they are equally famous for their wisdom, especially at Larisa, which is the native city of your friend Aristippus. And this is Gorgias' doing; for when he came there, the flower of the Aleuadae, among them your admirer Aristippus, and the other chiefs of the Thessalians, fell in love with his wisdom. And he has taught you the habit of answering questions in a grand and bold style, which becomes those who know, and is the style in which he himself answers all comers; and any Hellene who likes may ask him anything. How different is our lot! my dear Meno. Here at Athens there is a dearth of the commodity, and all wisdom seems to have emigrated from us to you. I am certain that if you were to ask any Athenian whether virtue was natural or acquired, he would laugh in your face, and say: "Stranger, you have far too good an opinion of me, if you think that I can answer your question. For I literally do not know what virtue is, and much less whether it is acquired by teaching or not." And I myself, Meno, living as I do in this region of poverty, am as poor as the rest of the world; and I confess with shame that I know literally nothing about virtue; and when I do not know the "quid" of anything how can I know the "quale"? How, if I knew nothing at all of Meno, could I tell if he was fair, or the opposite of fair; rich and noble, or the reverse of rich and noble? Do you think that I could?

Men. No, Indeed. But are you in earnest, Socrates, in saying that you do not know what virtue is? And am I to carry back this report of you to Thessaly?

Soc. Not only that, my dear boy, but you may say further that I have never known of any one else who did, in my judgment.

Men. Then you have never met Gorgias when he was at Athens?

Soc. Yes, I have.

Men. And did you not think that he knew?

Soc. I have not a good memory, Meno, and therefore I cannot now tell what I thought of him at the time. And I dare say that he did know, and that you know what he said: please, therefore, to remind me of what he said; or, if you would rather, tell me your own view; for I suspect that you and he think much alike.

Men. Very true.

9.4.2   NATURAL FITNESS

*"**MENO:** There are a great many other excellences [fitness's] too, and this makes it easy to say what excellence [fitness] is. For every task we undertake, there is, for each of us, the excellence [fitness] depends on our walk of life and our age, and I should imagine, Socrates, that by the same token there is for each of us the appropriate defect [unfitness] too."*

Plato, "Meno"
Robin Waterfield, "Plato: Meno and Other Dialogues"

In this section, an evolutionary concept is introduced. Plato has Meno assert that fitness is homo sapiens excellence – when a homo sapiens is fitted to their role within a society, or social ecosystem – that is excellence. Therefore, homo sapiens excellence can exist in many different patterns. However, Plato for the first time has Socrates assert there is only one pattern of homo sapiens excellence.

Plato then pivots from our species to another species organized socially like homo sapiens – bees. He then has Socrates ask Meno his "opinion" on "what it is to be a bee". Plato does so to reframe the discussion indirectly. He wants the audience to observe homo sapiens' overall nature just as they would bees in a forest.

In fact, Plato is really asking what do all homo sapiens have to do in every scenario Meno presented? Then Plato drops the word "imagine" into the dialogue. For that is the answer to his question. To become fitted to any role in our social ecosystem requires the adaptive trait of imagination. That is the single characteristic required to achieve Meno's many patterns of excellence.

This now introduces another evolutionary concept – adaption. Plato is asserting that imagination is required for homo sapiens to adapt and become fitted within any environment. Every homo sapiens in every society on the planet adapts in the same manner – with the help of imagination.

Soc. Then as he is not here, never mind him, and do you tell me: By the gods, Meno, be generous, and tell me what you say that virtue is; for I shall be truly delighted to find that I have been mistaken, and that you and Gorgias do really have this knowledge; although I have been just saying that I have never found anybody who had.

Men. There will be no difficulty, Socrates, in answering your question. Let us take first the virtue of a man-he should know how to administer the state, and in the administration of it to benefit his friends and harm his enemies; and he must also be careful not to suffer harm himself. A woman's virtue, if you wish to know about that, may also be easily described: her duty is to order her house, and keep what is indoors, and obey her husband. Every age, every condition of life, young or old, male or female, bond or free, has a different virtue: there are virtues numberless, and no lack of definitions of them; for virtue is relative to the actions and ages of each of us in all that we do. And the same may be said of vice, Socrates.

Soc. How fortunate I am, Meno! When I ask you for one virtue, you present me with a swarm of them, which are in your keeping. Suppose that I carry on the figure of the swarm, and ask of you, What is the nature of the bee? and you answer that there are many kinds of bees, and I reply: But do bees differ as bees, because there are many and different kinds of them; or are they not rather to be distinguished by some other quality, as for example beauty, size, or shape? How would you answer me?

Men. I should answer that bees do not differ from one another, as bees.

Soc. And if I went on to say: That is what I desire to know, Meno; tell me what is the quality in which they do not differ, but are all alike;-would you be able to answer?

Men. I should.

Soc. And so of the virtues, however many and different they may be, they have all a common nature which makes them virtues; and on this he who would answer the question, "What is virtue?" would do well to have his eye fixed: Do you understand?

Men. I am beginning to understand; but I do not as yet take hold of the question as I could wish.

Soc. When you say, Meno, that there is one virtue of a man, another of a woman, another of a child, and so on, does this apply only to virtue, or would you say the same of health, and size, and strength? Or is the nature of health always the same, whether in man or woman?

Men. I should say that health is the same, both in man and woman.

Soc. And is not this true of size and strength? If a woman is strong, she will be strong by reason of the same form and of the same strength subsisting in her which there is in the man. I mean to say that strength, as strength, whether of man or woman, is the same. Is there any difference?

Men. I think not.

Soc. And will not virtue, as virtue, be the same, whether in a child or in a grown-up person, in a woman or in a man?

Men. I cannot help feeling, Socrates, that this case is different from the others.

9.4.3   NATURAL PERFECTION

"SOCRATES: But they wouldn't be good [perfected] in the same way, I imagine, unless they had the same excellence [fitness].

Plato, "Meno"
Robin Waterfield, "Plato: Meno and Other Dialogues"

In this section, Plato introduces several evolutionary concepts. He first describes two human adaptive traits – justice and temperance. Justice is an adaptive trait in the form of behavior. It enables humans to maintain relative equilibrium in all human relationships and society. Aristotle covers this concept in detail in *Nichomachaen Ethics*.

Without relative justice human society would eventually collapse as sustained cooperation would be impossible. Temperance is also an adaptive trait in the form of behavior. Temperance is the act of restraining homo sapiens' natural instincts and impulses. It is presumed Plato envisions this within the legal structure of a city state. If temperance fails, justice is then required to remedy the injury done to others in society.

Plato then introduces the concept of perfection as being "good". He has Socrates state that the adaptive traits of justice and temperance are required to achieve perfection. To the ancient Greeks to develop a set of behavioral adaptations that fully realized your potential in any endeavor was perfection. To achieve basic competence at something was to be fit. To master that competency was to perfect it. This was homo sapiens perfection and was thought by the Greeks to be godlike. Aristotle expresses this more explicitly in the *Nichomachean Ethics*.

Plato points out that a homo sapiens cannot be perfected in the same way unless its form has the same fitness. This is logically correct. He also inserts the word "imagine" into the sentence. He places it deliberately between perfection and fitness. This is where imagination resides in the process of adapting from fitness to perfection. Here he literally gives you the pattern that produces excellence but does so subliminally. Plato then quickly redirects the reader to another red herring in the form of governing men.

Next Plato circles back to expand his list of adaptive traits that are required to achieve perfection. He now adds courage and wisdom to justice and temperance. These are known as Plato's four cardinal adaptive traits necessary for homo sapiens to achieve relative perfection in society. He then has Socrates indirectly plant the idea in the reader's mind that "one excellence" cuts across them all.

So, Plato means that imagination is necessary to create these virtues within homo sapiens. He then has Socrates state he can't "grasp" or imagine what the single excellence is that enables homo sapiens to adapt to any situation. Again he subtly gives you the answer within the sentence itself. It is imagination that enables homo sapiens to relatively perfect ourselves to any environment.

Soc. But why? Were you not saying that the virtue of a man was to order a state, and the virtue of a woman was to order a house?

Men. I did say so. Soc. And can either house or state or anything be well ordered without temperance and without justice? Men. Certainly not.

Soc. Then they who order a state or a house temperately or justly order them with temperance and justice? Men. Certainly.

Soc. Then both men and women, if they are to be good men and women, must have the same virtues of temperance and justice?

Men. True.

Soc. And can either a young man or an elder one be good, if they are intemperate and unjust?

Men. They cannot.

Soc. They must be temperate and just?

Men. Yes.

Soc. Then all men are good in the same way, and by participation in the same virtues?

Men. Such is the inference.

Soc. And they surely would not have been good in the same way, unless their virtue had been the same? Men. They would not.

Soc. Then now that the sameness of all virtue has been proven, try and remember what you and Gorgias say that virtue is.

Men. Will you have one definition of them all?

Soc. That is what I am seeking.

Men. If you want to have one definition of them all, I know not what to say, but that virtue is the power of governing mankind.

Soc. And does this definition of virtue include all virtue? Is virtue the same in a child and in a slave, Meno? Can the child govern his father, or the slave his master; and would he who governed be any longer a slave?

Men. I think not, Socrates.

Soc. No, indeed; there would be small reason in that. Yet once more, fair friend; according to you, virtue is "the power of governing"; but do you not add "justly and not unjustly"?

Men. Yes, Socrates; I agree there; for justice is virtue.

Soc. Would you say "virtue," Meno, or "a virtue"?

Men. What do you mean?

Soc. I mean as I might say about anything; that a round, for example, is "a figure" and not simply "figure," and I should adopt this mode of speaking, because there are other figures.

Men. Quite right; and that is just what I am saying about virtue-that there are other virtues as well as justice.

Soc. What are they? tell me the names of them, as I would tell you the names of the other figures if you asked me.

Men. Courage and temperance and wisdom and magnanimity are virtues; and there are many others.

Soc. Yes, Meno; and again we are in the same case: in searching after one virtue we have found many, though not in the same way as before; but we have been unable to find the common virtue which runs through them all.

Men. Why, Socrates, even now I am not able to follow you in the attempt to get at one common notion of virtue as of other things.

9.4.4   THE PATTERN OF EXCELLENCE

"SOCRATES: ...But come on, now: try to keep your promise and tell me what excellence is as a whole. Stop 'turning one into many'...leave excellence whole and intact, and tell me what it is. I've already supplied you with some models [patterns] to follow."

Plato, "Meno"
Robin Waterfield, "Plato: Meno and Other Dialogues"

In this section Plato addresses the importance of accurately defining a concept or pattern. He does so by having Socrates and Meno conduct a back-and-forth discussion of the concepts or patterns of "shape" (object) and "colours" (images). Socrates draws the reader's attention to the concept itself and its' various manifestations – a critical distinction in ancient Greek philosophy.

Plato demonstrates that Meno's attempt to define excellence by simply listing its' different manifestations is ineffective. It does not suffice as a full definition for the concept of excellence. In addition, Plato seeks to show how important precision in language is to philosophical discussions. Without such precision philosophers will be unable to "unconceal" the underlying pure concept or pattern of an object such as excellence.

Soc. No wonder; but I will try to get nearer if I can, for you know that all things have a common notion. Suppose now that some one asked you the question which I asked before: Meno, he would say, what is figure? And if you answered "roundness," he would reply to you, in my way of speaking, by asking whether you would say that roundness is "figure" or "a figure"; and you would answer "a figure."

Men. Certainly.

Soc. And for this reason-that there are other figures?

Men. Yes.

Soc. And if he proceeded to ask, What other figures are there? you would have told him.

Men. I should.

Soc. And if he similarly asked what colour is, and you answered whiteness, and the questioner rejoined, Would you say that whiteness is colour or a colour? you would reply, A colour, because there are other colours as well.

Men. I should.

Soc. And if he had said, Tell me what they are? -- you would have told him of other colours which are colours just as much as whiteness.

Men. Yes. **Soc**. And suppose that he were to pursue the matter in my way, he would say: Ever and anon we are landed in particulars, but this is not what I want; tell me then, since you call them by a common name, and say that they are all figures, even when opposed to one another, what is that common nature which you designate as figure-which contains straight as well as round, and is no more one than the other-that would be your mode of speaking?

Men. Yes.

Soc. And in speaking thus, you do not mean to say that the round is round any more than straight, or the straight any more straight than round?

Men. Certainly not.

Soc. You only assert that the round figure is not more a figure than the straight, or the straight than the round?

Men. Very true.

Soc. To what then do we give the name of figure? Try and answer. Suppose that when a person asked you this question either about figure or colour, you were to reply, Man, I do not understand what you want, or know what you are saying; he would look rather astonished and say: Do you not understand that I am looking for the "simile in multis"? And then he might put the question in another form: Mono, he might say, what is that "simile in multis" which you call figure, and which includes not only round and straight figures, but all? Could you not answer that question, Meno? I wish that you would try; the attempt will be good practice with a view to the answer about virtue.

Men. I would rather that you should answer, Socrates.

Soc. Shall I indulge you?

Men. By all means.

Soc. And then you will tell me about virtue?

Men. I will.

Soc. Then I must do my best, for there is a prize to be won.

Men. Certainly.

Soc. Well, I will try and explain to you what figure is. What do you say to this answer?-Figure is the only thing which always follows colour. Will you be satisfied with it, as I am sure that I should be, if you would let me have a similar definition of virtue?

Men. But, Socrates, it is such a simple answer.

Soc. Why simple?

Men. Because, according to you, figure is that which always follows colour.

Soc. Granted.

Men. But if a person were to say that he does not know what colour is, any more than what figure is-what sort of answer would you have given him?

Soc. I should have told him the truth. And if he were a philosopher of the eristic and antagonistic sort, I should say to him: You have my answer, and if I am wrong, your business is to take up the argument and refute me. But if we were friends, and were talking as you and I are now, I should reply in a milder strain and more in the dialectician's vein; that is to say, I should not only speak the truth, but I should make use of premisses which the person interrogated would be willing to admit. And this is the way in which I shall endeavour to approach you. You will acknowledge, will you not, that there is such a thing as an end, or termination, or extremity?-all which words use in the same sense, although I am aware that Prodicus might draw distinctions about them: but still you, I am sure, would speak of a thing as ended or terminated-that is all which I am saying-not anything very difficult.

Men. Yes, I should; and I believe that I understand your meaning.

Soc. And you would speak of a surface and also of a solid, as for example in geometry.

Men. Yes.

Soc. Well then, you are now in a condition to understand my definition of figure. I define figure to be that in which the solid ends; or, more concisely, the limit of solid.

Men. And now, Socrates, what is colour?

Soc. You are outrageous, Meno, in thus plaguing a poor old man to give you an answer, when you will not take the trouble of remembering what is Gorgias' definition of virtue.

Men. When you have told me what I ask, I will tell you, Socrates.

Soc. A man who was blindfolded has only to hear you talking, and he would know that you are a fair creature and have still many lovers.

Men. Why do you think so? Soc. Why, because you always speak in imperatives: like all beauties when they are in their prime, you are tyrannical; and also, as I suspect, you have found out that I have weakness for the fair, and therefore to humour you I must answer.

Men. Please do.

Soc. Would you like me to answer you after the manner of Gorgias, which is familiar to you?

Men. I should like nothing better.

Soc. Do not he and you and Empedocles say that there are certain effluences of existence?

Men. Certainly.

Soc. And passages into which and through which the effluences pass?

Men. Exactly.

Soc. And some of the effluences fit into the passages, and some of them are too small or too large?

Men. True.

Soc. And there is such a thing as sight?

Men. Yes.

Soc. And now, as Pindar says, "read my meaning" colour is an effluence of form, commensurate with sight, and palpable to sense.

Men. That, Socrates, appears to me to be an admirable answer.

Soc. Why, yes, because it happens to be one which you have been in the habit of hearing: and your wit will have discovered, I suspect, that you may explain in the same way the nature of sound and smell, and of many other similar phenomena.

Men. Quite true.

Soc. The answer, Meno, was in the orthodox solemn vein, and therefore was more acceptable to you than the other answer about figure.

Men. Yes.

Soc. And yet, O son of Alexidemus, I cannot help thinking that the other was the better; and I am sure that you would be of the same opinion, if you would only stay and be initiated, and were not compelled, as you said yesterday, to go away before the mysteries.

Men. But I will stay, Socrates, if you will give me many such answers.

Soc. Well then, for my own sake as well as for yours, I will do my very best; but I am afraid that I shall not be able to give you very many as good: and now, in your turn, you are to fulfil your promise, and tell me what virtue is in the universal; and do not make a singular into a plural, as the facetious say of those who break a thing, but deliver virtue to me whole and sound, and not broken into a number of pieces: I have given you the pattern.

9.4.5 PATTERN RECOGNITION

"MENO: There are people who think that bad things do them good, and then there are others who recognize that they do them harm."

Plato, "Meno"
Robin Waterfield, "Plato: Meno and Other Dialogues"

In this section, Plato describes the importance of pattern recognition. He has Socrates and Meno conduct a back-and-forth discussion about whether excellence is the ability to acquire good or desirable things. The initial assumption is that homo sapiens only want good things.

However, they then agree that some people want bad things (e.g., heroin, cigarettes, etc.). This is because some people lack the adaptive trait wisdom (i.e., ability to discern) to recognize bad things. They think erroneously that the bad things are good things. So, pattern recognition is critical in discerning good things from bad things.

Next Plato has Socrates and Meno discuss the importance of some aspects of excellence (e.g., justice, temperance, etc.) being present during the acquisition of good things. This is equivalent to saying that if you won an Olympic medal by cheating (i.e., unjustly) that is not excellence. But winning the Olympic medal by attaining perfection in your competency and beating your opponent justly is excellence. Alternately, to not acquire a desired thing by resisting the temptation to act unjustly is also excellence. To do so in either case was considered to be godlike. This is why the ancient Greeks conducted the Olympic Games together with many religious ceremonies.

Unfortunately, Meno's logic was circular. He stated that excellence was justice, but he already agreed that Justice was a part of excellence. A thing cannot be defined by a part of what comprises it. Socrates earlier pointed out this logical fallacy when Meno presented a plurality of excellences. This new definition is again not sufficient.

Men. Well then, Socrates, virtue, as I take it, is when he, who desires the honourable, is able to provide it for himself; so the poet says, and I say too- Virtue is the desire of things honourable and the power of attaining them.

Soc. And does he who desires the honourable also desire the good?

Men. Certainly.

Soc. Then are there some who desire the evil and others who desire the good? Do not all men, my dear sir, desire good?

Men. I think not.

Soc. There are some who desire evil?

Men. Yes.

Soc. Do you mean that they think the evils which they desire, to be good; or do they know that they are evil and yet desire them?

Men. Both, I think.

Soc. And do you really imagine, Meno, that a man knows evils to be evils and desires them notwithstanding?

Men. Certainly I do.

Soc. And desire is of possession? Men. Yes, of possession.

Soc. And does he think that the evils will do good to him who possesses them, or does he know that they will do him harm?

Men. There are some who think that the evils will do them good, and others who know that they will do them harm.

Soc. And, in your opinion, do those who think that they will do them good know that they are evils?

Men. Certainly not.

Soc. Is it not obvious that those who are ignorant of their nature do not desire them; but they desire what they suppose to be goods although they are really evils; and if they are mistaken and suppose the evils to be good they really desire goods?

Men. Yes, in that case.

Soc. Well, and do those who, as you say, desire evils, and think that evils are hurtful to the possessor of them, know that they will be hurt by them?

Men. They must know it.

Soc. And must they not suppose that those who are hurt are miserable in proportion to the hurt which is inflicted upon them? Men. How can it be otherwise?

Soc. But are not the miserable ill-fated?

Men. Yes, indeed.

Soc. And does any one desire to be miserable and ill-fated?

Men. I should say not, Socrates.

Soc. But if there is no one who desires to be miserable, there is no one, Meno, who desires evil; for what is misery but the desire and possession of evil?

Men. That appears to be the truth, Socrates, and I admit that nobody desires evil.

Soc. And yet, were you not saying just now that virtue is the desire and power of attaining good?

Men. Yes, I did say so.

Soc. But if this be affirmed, then the desire of good is common to all, and one man is no better than another in that respect?

Men. True. Soc. And if one man is not better than another in desiring good, he must be better in the power of attaining it?

Men. Exactly.

Soc. Then, according to your definition, virtue would appear to be the power of attaining good?

Men. I entirely approve, Socrates, of the manner in which you now view this matter.

Soc. Then let us see whether what you say is true from another point of view; for very likely you may be right:-You affirm virtue to be the power of attaining goods?

Men. Yes.

Soc. And the goods which mean are such as health and wealth and the possession of gold and silver, and having office and honour in the state-those are what you would call goods?

Men. Yes, I should include all those.

Soc. Then, according to Meno, who is the hereditary friend of the great king, virtue is the power of getting silver and gold; and would you add that they must be gained piously, justly, or do you deem this to be of no consequence? And is any mode of acquisition, even if unjust and dishonest, equally to be deemed virtue?

Men. Not virtue, Socrates, but vice.

Soc. Then justice or temperance or holiness, or some other part of virtue, as would appear, must accompany the acquisition, and without them the mere acquisition of good will not be virtue. Men. Why, how can there be virtue without these?

Soc. And the non-acquisition of gold and silver in a dishonest manner for oneself or another, or in other words the want of them, may be equally virtue?

Men. True.

Soc. Then the acquisition of such goods is no more virtue than the non-acquisition and want of them, but whatever is accompanied by justice or honesty is virtue, and whatever is devoid of justice is vice. Men. It cannot be otherwise, in my judgment.

Soc. And were we not saying just now that justice, temperance, and the like, were each of them a part of virtue?

Men. Yes.

Soc. And so, Meno, this is the way in which you mock me.

Men. Why do you say that, Socrates?

Soc. Why, because I asked you to deliver virtue into my hands whole and unbroken, and I gave you a pattern according to which you were to frame your answer; and you have forgotten already, and tell me that virtue is the power of attaining good justly, or with justice; and justice you acknowledge to be a part of virtue.

Men. Yes.

Soc. Then it follows from your own admissions, that virtue is doing what you do with a part of virtue; for justice and the like are said by you to be parts of virtue.

Men. What of that?

Soc. What of that! Why, did not I ask you to tell me the nature of virtue as a whole? And you are very far from telling me this; but declare every action to be virtue which is done with a part of virtue; as though you had told me and I must already know the whole of virtue, and this too when frittered away into little pieces. And, therefore, my dear I fear that I must begin again and repeat the same question: What is virtue? for otherwise, I can only say, that every action done with a part of virtue is virtue; what else is the meaning of saying that every action done with justice is virtue? Ought I not to ask the question over again; for can any one who does not know virtue know a part of virtue?

Men. No; I do not say that he can.

Soc. Do you remember how, in the example of figure, we rejected any answer given in terms which were as yet unexplained or unadmitted?

Men. Yes, Socrates; and we were quite right in doing so.

Soc. But then, my friend, do not suppose that we can explain to any one the nature of virtue as a whole through some unexplained portion of virtue, or anything at all in that fashion; we should only have to ask over again the old question, What is virtue? Am I not right?

Men. I believe that you are.

Soc. Then begin again, and answer me, What, according to you and your friend Gorgias, is the definition of virtue?

9.4.6   MENO'S PARADOX

"__SOCRATES:__ I see what you're getting at, Meno...The claim is that it's impossible for a man to search either for what he knows or for what he doesn't know: he wouldn't be searching for what he knows, since he knows it and that makes the search unnecessary, and he can't search for what he doesn't know either, since he doesn't even know what it is he's going to search for."

Plato, "Meno"
Robin Waterfield, "Plato: Meno and Other Dialogues"

In this section, Plato has the characters summarize the progress they have made. Meno confesses that he no longer knows what excellence is and that he previously didn't know what he didn't know. Meno now knows what he doesn't know about excellence.

In his frustration, Meno compares Socrates to a torpedo fish. It seems like a comedic aside, but it is in truth not. Plato is demonstrating to the reader a form of pattern matching. Meno matches one element of Socrates' behavior to one element of something else's behavior. Menos' brain did this spontaneously at the subconscious level due to his frustration. This is an important data point that Plato will build on in the next section.

Plato then introduces through Meno a new problem in the enquiry into excellence. Meno didn't know that he didn't know what excellence was. But how could he have remedied this – if you already know something you have no need to **search** for it. If you don't know something, how can you search for something you don't know you don't know? How would you even know that you found it when you don't know what it is? This is what is known as "Meno's Paradox". However, Socrates does not think Meno's argument is sound. He presents new information that resolves the apparent paradox.

Men. O Socrates, I used to be told, before I knew you, that you were always doubting yourself and making others doubt; and now you are casting your spells over me, and I am simply getting bewitched and enchanted, and am at my wits' end. And if I may venture to make a jest upon you, you seem to me both in your appearance and in your power over others to be very like the flat torpedo fish, who torpifies those who come near him and touch him, as you have now torpified me, I think. For my soul and my tongue are really torpid, and I do not know how to answer you; and though I have been delivered of an infinite variety of speeches about virtue before now, and to many persons-and very good ones they were, as I thought-at this moment I cannot even say what virtue is. And I think that. you are very wise in not voyaging and going away from home, for if you did in other places as do in Athens, you would be cast into prison as a magician.

Soc. You are a rogue, Meno, and had all but caught me.

Men. What do you mean, Socrates?

Soc. I can tell why you made a simile about me.

Men. Why?

Soc. In order that I might make another simile about you. For I know that all pretty young gentlemen like to have pretty similes made about them-as well they may-but I shall not return the compliment. As to my being a torpedo, if the torpedo is torpid as well as the cause of torpidity in others, then indeed I am a torpedo, but not otherwise; for I perplex others, not because I am clear, but because I am utterly perplexed myself. And now I know not what virtue is, and you seem to be in the same case, although you did once perhaps know before you touched me. However, I have no objection to join with you in the enquiry.

Men. And how will you enquire, Socrates, into that which you do not know? What will you put forth as the subject of enquiry? And if you find what you want, how will you ever know that this is the thing which you did not know?

Soc. I know, Meno, what you mean; but just see what a tiresome dispute you are introducing. You argue that man cannot enquire either about that which he knows, or about that which he does not know; for if he knows, he has no need to enquire; and if not, he cannot; for he does not know the, very subject about which he is to enquire.

Men. Well, Socrates, and is not the argument sound?

Soc. I think not.

Men. Why not?

Soc. I will tell you why: I have heard from certain wise men and women who spoke of things divine that-

Men. What did they say?

Soc. They spoke of a glorious truth, as I conceive.

Men. What was it? and who were they?

Soc. Some of them were priests and priestesses, who had studied how they might be able to give a reason of their profession: there, have been poets also, who spoke of these things by inspiration, like Pindar, and many others who were inspired. And they say-mark, now, and see whether their words are true-they say that the soul of man is immortal, and at one time has an end, which is termed dying, and at another time is born again, but is never destroyed. And the moral is, that a man ought to live always in perfect holiness. "For in the ninth year Persephone sends the souls of those from whom she has received the penalty of ancient crime back again from beneath into the light of the sun above, and these are they who become noble kings and mighty men and great in wisdom and are called saintly heroes in after ages." The soul, then, as being immortal, and having been born again many times, rand having seen all things that exist, whether in this world or in the world below, has knowledge of them all; and it is no wonder that she should be able to call to remembrance all that she ever knew about virtue, and about everything; for as all nature is akin, and the soul has learned all things; there is no difficulty in her eliciting or as men say learning, out of a single recollection -all the rest, if a man is strenuous and does not faint; for all enquiry and all learning is but recollection. And therefore we ought not to listen to this sophistical argument about the impossibility of enquiry: for it will make us idle; and is sweet only to the sluggard; but the other saying will make us active and inquisitive. In that confiding, I will gladly enquire with you into the nature of virtue.

Men. Yes, Socrates; but what do you mean by saying that we do not learn, and that what we call learning is only a process of recollection? Can you teach me how this is?

Soc. I told you, Meno, just now that you were a rogue, and now you ask whether I can teach you, when I am saying that there is no teaching, but only recollection; and thus you imagine that you will involve me in a contradiction.

Men. Indeed, Socrates, I protest that I had no such intention. I only asked the question from habit; but if you can prove to me that what you say is true, I wish that you would.

Soc. It will be no easy matter, but I will try to please you to the utmost of my power. Suppose that you call one of your numerous attendants, that I may demonstrate on him.

9.4.7   PATTERN MATCHING

"SOCRATES: Given that the human soul [mind] is immortal and has been reincarnated many times, and has therefore seen things here on earth...For all nature [evolution] is akin and the soul [natural adaptation of the human brain] has learnt everything, there's nothing to stop a man from recovering [pattern-matching] everything else by himself, once he has remembered – or 'learnt'...just one thing [a single concept in a pattern]...The point is that the search, the process of learning, is in fact nothing but recollection [pattern-matching using our imagination]."

Plato, "Meno"
Robin Waterfield, "Plato: Meno and Other Dialogues"

In this section Plato proposes the solution to "Meno's Paradox" – the process of "recollection". Recollection is really just pattern-matching. As Malcolm Gladwell states in his book *Blink: The Power of Thinking Without Thinking*:

"The part of our brain that leaps to conclusions is called the adaptive unconscious...This new notion of the adaptive unconscious is thought of, instead, as a kind of giant computer that quickly and quietly processes a lot of data we need in order to keep functioning as human beings."

In the "human soul is immortal" concept Plato is describing the human mind and how it functions. He's not describing it literally, but how it practically works. Basically, the homo sapiens brain is a pattern matching machine. Amazon's founder and chief executive officer, Jeff Bezos, literally made this same observation:

"The human brain is an incredible pattern-matching machine."

We must think back to before homo sapiens created language and civilization. In our primitive past we operated in a natural environment characterized by a set of repetitive patterns (i.e., lunar, solar, weather, seasons, predator, prey, etc.). These patterns have not changed fundamentally since our species' pre-conscious existence. Our pattern matching adaptation enabled us to perceive and then anticipate persistent natural patterns. This dramatically enhances a species' chance of survival.

In fact, this very dialectic is constructed to manipulate your brain's subconscious and automatic pattern matching instinct. Plato is slipping patterns into your subconscious mind throughout the dialectic. He wants your brain to eventually pattern match consciously on a pattern he already slipped into your subconscious. You will think it is your idea, but really it was Plato's which he implanted in your mind earlier in the dialectic. That is the reason why he places two human characters, Socrates and Meno, in front of himself as a "shiny object" to distract you.

In a sense, the human species isn't really learning anything new because "all nature is akin". Our instincts and brain are designed to operate in these natural patterns. That is what Plato means by recollecting – it doesn't require learning or teaching as we understand it. The patterns our ancestors matched three million years ago, and the ones we match today are still the same – if a slightly different variation. Meno's pattern matching of Socrates to the torpedo fish was a literal example of this phenomenon.

Meno now "knows" that about Socrates. However, Meno didn't know he didn't know this about Socrates prior to engaging in this dialogue. But Meno did know the pattern of the torpedo fish prior to the dialogue with Socrates. So, did Meno truly learn something completely new? At the conceptual level, no. He simply matched a conceptual pattern he already knew onto another "thing".

Meno literally recollected the conceptual pattern of the torpedo fish. All it required was a little imagination to match it with Socrates. This is why Socrates says pattern matching can't be taught. Why would you need to be taught something you can already do? It is the fact that Meno is ignorant of how his mind works that he requires teaching. What is someone teaching you other than what was discovered through pattern-matching and imagination? Logically that means you could also have discovered it for yourself.

This is why the "Meno Paradox" isn't a paradox at all. How does a human search for something they don't know they don't know? You do it every day. Often, you identify one piece of data in a pattern. Then, you work out the rest of the pattern bit by bit. Often that pattern leads to you discovering something unexpected. Many important discoveries in history have been found accidentally through this process of pattern matching. It just requires persistence and imagination to work out the patterns present in nature. This is the process by which homo sapiens has generated all existing knowledge.

Men. Certainly. Come hither, boy.

Soc. He is Greek, and speaks Greek, does he not?

Men. Yes, indeed; he was born in the house.

Soc. Attend now to the questions which I ask him, and observe whether he learns of me or only remembers.

Men. I will. Soc. Tell me, boy, do you know that a figure like this is a square? Boy. I do.

Soc. And you know that a square figure has these four lines equal?

Boy. Certainly.

Soc. And these lines which I have drawn through the middle of the square are also equal?

Boy. Yes.

Soc. A square may be of any size?

Boy. Certainly.

Soc. And if one side of the figure be of two feet, and the other side be of two feet, how much will the whole be? Let me explain: if in one direction the space was of two feet, and in other direction of one foot, the whole would be of two feet taken once?

Boy. Yes.

Soc. But since this side is also of two feet, there are twice two feet?

Boy. There are.

Soc. Then the square is of twice two feet?

Boy. Yes.

Soc. And how many are twice two feet? count and tell me.

Boy. Four, Socrates.

Soc. And might there not be another square twice as large as this, and having like this the lines equal?

Boy. Yes.

Soc. And of how many feet will that be?

Boy. Of eight feet.

Soc. And now try and tell me the length of the line which forms the side of that double square: this is two feet-what will that be?

Boy. Clearly, Socrates, it will be double.

Soc. Do you observe, Meno, that I am not teaching the boy anything, but only asking him questions; and now he fancies that he knows how long a line is necessary in order to produce a figure of eight square feet; does he not?

Men. Yes.

Soc. And does he really know?

Men. Certainly not.

Soc. He only guesses that because the square is double, the line is double.

Men. True.

Soc. Observe him while he recalls the steps in regular order. (To the Boy.) Tell me, boy, do you assert that a double space comes from a double line? Remember that I am not speaking of an oblong, but of a figure equal every way, and twice the size of this-that is to say of eight feet; and I want to know whether you still say that a double square comes from double line?

Boy. Yes.

Soc. But does not this line become doubled if we add another such line here?

Boy. Certainly.

Soc. And four such lines will make a space containing eight feet?

Boy. Yes.

Soc. Let us describe such a figure: Would you not say that this is the figure of eight feet?

Boy. Yes.

Soc. And are there not these four divisions in the figure, each of which is equal to the figure of four feet?

Boy. True.

Soc. And is not that four times four?

Boy. Certainly.

Soc. And four times is not double?

Boy. No, indeed. Soc. But how much?

Boy. Four times as much.

Soc. Therefore the double line, boy, has given a space, not twice, but four times as much.

Boy. True.

Soc. Four times four are sixteen-are they not?

Boy. Yes.

Soc. What line would give you a space of eight feet, as this gives one of sixteen feet;-do you see?

Boy. Yes.

Soc. And the space of four feet is made from this half line?

Boy. Yes.

Soc. Good; and is not a space of eight feet twice the size of this, and half the size of the other?

Boy. Certainly.

Soc. Such a space, then, will be made out of a line greater than this one, and less than that one?

Boy. Yes; I think so.

Soc. Very good; I like to hear you say what you think. And now tell me, is not this a line of two feet and that of four?

Boy. Yes.

Soc. Then the line which forms the side of eight feet ought to be more than this line of two feet, and less than the other of four feet?

Boy. It ought.

Soc. Try and see if you can tell me how much it will be.

Boy. Three feet.

Soc. Then if we add a half to this line of two, that will be the line of three. Here are two and there is one; and on the other side, here are two also and there is one: and that makes the figure of which you speak?

Boy. Yes.

Soc. But if there are three feet this way and three feet that way, the whole space will be three times three feet?

Boy. That is evident.

Soc. And how much are three times three feet?

Boy. Nine.

Soc. And how much is the double of four?

Boy. Eight.

Soc. Then the figure of eight is not made out of a of three?

Boy. No.

Soc. But from what line?-tell me exactly; and if you would rather not reckon, try and show me the line.

Boy. Indeed, Socrates, I do not know.

Soc. Do you see, Meno, what advances he has made in his power of recollection? He did not know at first, and he does not know now, what is the side of a figure of eight feet: but then he thought that he knew, and answered confidently as if he knew, and had no difficulty; now he has a difficulty, and neither knows nor fancies that he knows.

Men. True.

Soc. Is he not better off in knowing his ignorance?

Men. I think that he is.

Soc. If we have made him doubt, and given him the "torpedo's shock," have we done him any harm?

Men. I think not.

Soc. We have certainly, as would seem, assisted him in some degree to the discovery of the truth; and now he will wish to remedy his ignorance, but then he would have been ready to tell all the world again and again that the double space should have a double side.

Men. True.

Soc. But do you suppose that he would ever have enquired into or learned what he fancied that he knew, though he was really ignorant of it, until he had fallen into perplexity under the idea that he did not know, and had desired to know?

Men. I think not, Socrates.

Soc. Then he was the better for the torpedo's touch?

Men. I think so.

9.4.8   KNOWLEDGE GENERATION

"SOCRATES: So someone who doesn't know about whatever it is that he doesn't know has true beliefs inside him about these things he doesn't know...if he were to be repeatedly asked the same questions in a number

of different ways, he'd certainly end up with knowledge of these matters that is as good and as accurate as anyone's."

Plato, "Meno"
Robin Waterfield, "Plato: Meno and Other Dialogues"

Socrates initiates the process of discovery with the Slave. Plato selects a geometrical problem as it has a mathematically provable solution. This makes the answer to the problem quantitative or objective rather than qualitative or subjective. It has a definite answer which allows for no opinion such as: 1 + 1 = 2. If you believe it equals three, that is an opinion – which can be false or factually inaccurate.

The Slave initially forms a false opinion, but Socrates quickly points out it is inaccurate. By the end of the section the Slave is aware he doesn't know the answer. The Slave's inaccurate opinion has been removed and he knows what he doesn't know. However, the Slave wants is determined to find the answer. This is the starting point to acquire wisdom. As the Slave has generated new personal knowledge, he now knows he doesn't know how to produce the solution.

Plato then has Socrates repeat the exercise. He tells Meno he will not teach the Slave anything, but only ask questions. The Slave is then asked another series of questions. This time comes to the right answer on his own – without being explicitly told or taught the answer. Plato provides a supernatural answer to how this was accomplished. This is consistent with his ancient Greek worldview. But Plato does give the correct literal answer as to how the Slave discovered the answer. In Robin Waterfield's *Plato: Meno and Other Dialogues* Socrates states:

"SOCRATES: So if the truth of things is always in our souls [mind], the soul [mind] must be immortal [genetically], and this means that if there's something you happen to know at the moment...you can confidently try to search [imagine] for it and recall it. Yes?"

Plato's soul is really the human mind. So, Plato is saying the truth of all things can be discovered through the homo sapiens mind. That is how all homo sapiens knowledge has been generated. At one point everything we consider scientific knowledge was unknown. Homo sapiens didn't know we didn't know it. Afterwards we knew we didn't know it and were attempting to figure it out. We did so through pattern-matching and imagination. What is knowledge? Knowledge is information about the patterns in nature that have proved to be consistently true. Knowledge is a combination of information and patterns which can be reliably built on to discover new information and patterns in nature.

Soc. Mark now the farther development. I shall only ask him, and not teach him, and he shall share the enquiry with me: and do you watch and see if you find me telling or explaining anything to him, instead of eliciting his opinion. Tell me, boy, is not this a square of four feet which I have drawn?

Boy. Yes.

Soc. And now I add another square equal to the former one?

Boy. Yes.

Soc. And a third, which is equal to either of them?

Boy. Yes.

Soc. Suppose that we fill up the vacant corner?

Boy. Very good.

Soc. Here, then, there are four equal spaces?

Boy. Yes.

Soc. And how many times larger is this space than this other?

Boy. Four times.

Soc. But it ought to have been twice only, as you will remember.

Boy. True.

Soc. And does not this line, reaching from corner to corner, bisect each of these spaces?

Boy. Yes.

Soc. And are there not here four equal lines which contain this space?

Boy. There are.

Soc. Look and see how much this space is.

Boy. I do not understand.

Soc. Has not each interior line cut off half of the four spaces?

Boy. Yes.

Soc. And how many spaces are there in this section?

Boy. Four.

Soc. And how many in this?

Boy. Two.

Soc. And four is how many times two?

Boy. Twice.

Soc. And this space is of how many feet?

Boy. Of eight feet.

Soc. And from what line do you get this figure?

Boy. From this.

Soc. That is, from the line which extends from corner to corner of the figure of four feet?

Boy. Yes.

Soc. And that is the line which the learned call the diagonal. And if this is the proper name, then you, Meno's slave, are prepared to affirm that the double space is the square of the diagonal?

Boy. Certainly, Socrates.

Soc. What do you say of him, Meno? Were not all these answers given out of his own head?

Men. Yes, they were all his own.

Soc. And yet, as we were just now saying, he did not know?

Men. True. Soc. But still he had in him those notions of his-had he not?

Men. Yes.

Soc. Then he who does not know may still have true notions of that which he does not know?

Men. He has.

Soc. And at present these notions have just been stirred up in him, as in a dream; but if he were frequently asked the same questions, in different forms, he would know as well as any one at last?

Men. I dare say.

Soc. Without any one teaching him he will recover his knowledge for himself, if he is only asked questions?

Men. Yes.

Soc. And this spontaneous recovery of knowledge in him is recollection?

Men. True.

Soc. And this knowledge which he now has must he not either have acquired or always possessed?

Men. Yes.

Soc. But if he always possessed this knowledge he would always have known; or if he has acquired the knowledge he could not have acquired it in this life, unless he has been taught geometry; for he may be made to do the same with all geometry and every other branch of knowledge. Now, has any one ever taught him all this? You must know about him, if, as you say, he was born and bred in your house.

Men. And I am certain that no one ever did teach him.

Soc. And yet he has the knowledge?

Men. The fact, Socrates, is undeniable.

Soc. But if he did not acquire the knowledge in this life, then he must have had and learned it at some other time?

Men. Clearly he must.

Soc. Which must have been the time when he was not a man?

Men. Yes.

Soc. And if there have been always true thoughts in him, both at the time when he was and was not a man, which only need to be awakened into knowledge by putting questions to him, his soul must have always possessed this knowledge, for he always either was or was not a man?

Men. Obviously.

Soc. And if the truth of all things always existed in the soul, then the soul is immortal. Wherefore be of good cheer, and try to recollect what you do not know, or rather what you do not remember.

Men. I feel, somehow, that I like what you are saying.

Soc. And I, Meno, like what I am saying. Some things I have said of which I am not altogether confident. But that we shall be better and braver and less helpless if we think that we ought to enquire, than we should have been if we indulged in the idle fancy that there was no knowing and no use in seeking to know what we do not know;-that is a theme upon which I am ready to fight, in word and deed, to the utmost of my power.

Men. There again, Socrates, your words seem to me excellent.

Soc. Then, as we are agreed that a man should enquire about that which he does not know, shall you and I make an effort to enquire together into the nature of virtue?

Men. By all means, Socrates. And yet I would much rather return to my original question, Whether in seeking to acquire virtue we should regard it as a thing to be taught, or as a gift of nature, or as coming to men in some other way?

9.4.9   A MENTAL QUALITY

"SOCRATES: Since we don't know what it is or what sort of thing it is, let's look into the question of whether or not it's teachable by making the following assumption: what sort of mental quality would it have to be or not to be to be teachable?"

> Plato, "Meno"
> Robin Waterfield, "Plato: Meno and Other Dialogues"

Plato then has Socrates circle the conversation back to the search for the definition of excellence. Socrates asserts that it is a mental quality. The question is whether this mental quality is similar to knowledge. For we know that knowledge can be taught. If this mental quality can be taught, then it is a form of knowledge. If it is a form of knowledge, then logically it must be possible to teach excellence.

Socrates and Meno then agree that this specific mental quality is good for humans. They then compare this mental quality to other things that are good for people – strength, health, justice, courage, cleverness, memory, etc. Sometimes these things can harm people if a person does not utilize them effectively. In the end they agree that exercising this mental quality with persistence leads to excellence when guided by knowledge. The opposite of excellence is when this mental quality is guided by ignorance.

Soc. Had I the command of you as well as of myself, Meno, I would not have enquired whether virtue is given by instruction or not, until we had first ascertained "what it is." But as you think only of controlling me who am your slave, and never of controlling yourself,-such being your notion of freedom, I must yield to you, for you are irresistible. And therefore I have now to enquire into the qualities of a thing of which I do not as yet know the nature. At any rate, will you condescend a little, and allow the question "Whether virtue is given by instruction, or in any other way," to be argued upon hypothesis? As the geometrician, when he is asked whether a certain triangle is capable being inscribed in a certain circle, will reply: "I cannot tell you as yet; but I will offer a hypothesis which may assist us in forming a conclusion: If the figure be such that when you have produced a given side of it, the given area of the triangle falls short by an area corresponding to the part produced, then one consequence follows, and if this is impossible then some other; and therefore I wish to assume a hypothesis before I tell you whether this triangle is capable of being inscribed in the circle":-that is a geometrical hypothesis. And we too, as we know not the nature and -qualities of virtue, must ask, whether virtue is or not taught, under a hypothesis: as thus, if virtue is of such a class of mental goods, will it be taught or not? Let the first hypothesis be-that virtue is or is not knowledge,-in that case will it be taught or not? or, as we were just now saying, remembered"? For there is no use in disputing about the name. But is virtue taught or not? or rather, does not everyone see that knowledge alone is taught?

Men. I agree.

Soc. Then if virtue is knowledge, virtue will be taught?

Men. Certainly.

Soc. Then now we have made a quick end of this question: if virtue is of such a nature, it will be taught; and if not, not?

Men. Certainly.

Soc. The next question is, whether virtue is knowledge or of another species?

Men. Yes, that appears to be the -question which comes next in order.

Soc. Do we not say that virtue is a good?-This is a hypothesis which is not set aside.

Men. Certainly.

Soc. Now, if there be any sort-of good which is distinct from knowledge, virtue may be that good; but if knowledge embraces all good, then we shall be right in think in that virtue is knowledge?

Men. True.

Soc. And virtue makes us good?

Men. Yes.

Soc. And if we are good, then we are profitable; for all good things are profitable?

Men. Yes.

Soc. Then virtue is profitable?

Men. That is the only inference.

Soc. Then now let us see what are the things which severally profit us. Health and strength, and beauty and wealth-these, and the like of these, we call profitable?

Men. True.

Soc. And yet these things may also sometimes do us harm: would you not think so?

Men. Yes. Soc. And what is the guiding principle which makes them profitable or the reverse? Are they not profitable when they are rightly used, and hurtful when they are not rightly used?

Men. Certainly.

Soc. Next, let us consider the goods of the soul: they are temperance, justice, courage, quickness of apprehension, memory, magnanimity, and the like?

Men. Surely.

Soc. And such of these as are not knowledge, but of another sort, are sometimes profitable and sometimes hurtful; as, for example, courage wanting prudence, which is only a sort of confidence? When a man has no sense he is harmed by courage, but when he has sense he is profited?

Men. True.

Soc. And the same may be said of temperance and quickness of apprehension; whatever things are learned or done with sense are profitable, but when done without sense they are hurtful?

Men. Very true.

Soc. And in general, all that the attempts or endures, when under the guidance of wisdom, ends in happiness; but when she is under the guidance of folly, in the opposite?

Men. That appears to be true.

9.4.10 IS EXCELLENCE KNOWLEDGE?

"<u>SOCRATES:</u> So we're saying that excellence is knowledge – either the whole of knowledge or some part of it aren't we?"

Plato, "Meno"
Robin Waterfield, "Plato: Meno and Other Dialogues"

In this section Socrates states that you need knowledge to effectively exercise this mental quality. So, excellence must be knowledge. If excellence is knowledge, then it isn't genetic. This makes excellence teachable since knowledge can be taught. But Socrates has a strong doubt – he has never heard of any teachers or students of the mental quality which is excellence.

Soc. If then virtue is a quality of the soul, and is admitted to be profitable, it must be wisdom or prudence, since none of the things of the soul are either profitable or hurtful in themselves, but they are all made profitable or hurtful by the addition of wisdom or of folly; and therefore and therefore if virtue is profitable, virtue must be a sort of wisdom or prudence?

Men. I quite agree.

Soc. And the other goods, such as wealth and the like, of which we were just now saying that they are sometimes good and sometimes evil, do not they also become profitable or hurtful, accordingly as the soul guides and uses them rightly or wrongly; just as the things of the soul herself are benefited when under the guidance of wisdom and harmed by folly?

Men. True.

Soc. And the wise soul guides them rightly, and the foolish soul wrongly.

Men. Yes.

Soc. And is not this universally true of human nature? All other things hang upon the soul, and the things of the soul herself hang upon wisdom, if they are to be good; and so wisdom is inferred to be that which profits-and virtue, as we say, is profitable?

Men. Certainly.

Soc. And thus we arrive at the conclusion that virtue is either wholly or partly wisdom?

Men. I think that what you are saying, Socrates, is very true.

Soc. But if this is true, then the good are not by nature good?

Men. I think not.

Soc. If they had been, there would assuredly have been discerners of characters among us who would have known our future great men; and on their showing we should have adopted them, and when we had got them, we should have kept them in the citadel out of the way of harm, and set a stamp upon them far rather than upon a piece of gold, in order that no one might tamper with them; and when they grew up they would have been useful to the state?

Men. Yes, Socrates, that would have been the right way.

Soc. But if the good are not by nature good, are they made good by instruction?

Men. There appears to be no other alternative, Socrates. On the supposition that virtue is knowledge, there can be no doubt that virtue is taught.

Soc. Yes, indeed; but what if the supposition is erroneous?

Men. I certainly thought just now that we were right.

Soc. Yes, Meno; but a principle which has any soundness should stand firm not only just now, but always.

Men. Well; and why are you so slow of heart to believe that knowledge is virtue?

Soc. I will try and tell you why, Meno. I do not retract the assertion that if virtue is knowledge it may be taught; but I fear that I have some reason in doubting whether virtue is knowledge: for consider now. and say whether virtue, and not only virtue but anything that is taught, must not have teachers and disciples?

Men. Surely.

Soc. And conversely, may not the art of which neither teachers nor disciples exist be assumed to be incapable of being taught?

Men. True; but do you think that there are no teachers of virtue?

9.4.11   ARE THEIR TEACHERS OF EXCELLENCE?

"SOCRATES: So Anytus is typical of the kind of person with whom one ought to try to see whether or not there are any teachers of excellence, and if so who they are."

Plato, "Meno"
Robin Waterfield, "Plato: Meno and Other Dialogues"

In this section Plato has Socrates joined by a new character, Anytus. They discuss whether the mental quality that is excellence can be taught. However, this discussion is a contradiction in terms. Pattern-matching doesn't need to be taught since you are born able to do it. Plato is intentionally walking us through this false thinking in order eliminate these concepts from our enquiry. It is part of determining what a concept is not. In addition, Plato uses the discussion to subtly give the reader a pattern for false belief.

Soc. I have certainly often enquired whether there were any, and taken great pains to find them, and have never succeeded; and many have assisted me in the search, and they were the persons whom I thought the most likely to know. Here at the moment when he is wanted we fortunately have sitting by us Anytus, the very person of whom we should make enquiry; to him then let us repair. In the first Place, he is the son of a wealthy and wise father, Anthemion, who acquired his wealth, not by accident or gift, like Ismenias the Theban (who has recently made himself as rich as Polycrates), but by his own skill and industry, and who is a well-conditioned, conditioned, modest man, not insolent, or over-bearing, or annoying; moreover, this son of his has received a good education, as the Athenian people certainly appear to think, for they choose him to fill the highest offices. And these are the sort of men from whom you are likely to learn whether there are any teachers of virtue, and who they are. Please, Anytus, to help me and your friend Meno in answering our question, Who are the teachers? Consider the matter thus: If we wanted Meno to be a good physician, to whom should we send him? Should we not send him to the physicians?

Any. Certainly.

Soc. Or if we wanted him to be a good cobbler, should we not send him to the cobblers?

Any. Yes.

Soc. And so forth?

Any. Yes.

Soc. Let me trouble you with one more question. When we say that we should be right in sending him to the physicians if we wanted him to be a physician, do we mean that we should be right in sending him to those who profess the art, rather than to those who do not, and to those who demand payment for teaching the art, and profess to teach it to any one who will come and learn? And if these were our reasons, should we not be right in sending him?

Any. Yes.

Soc. And might not the same be said of flute-playing, and of the other arts? Would a man who wanted to make another a flute-player refuse to send him to those who profess to teach the art for money, and be plaguing other persons to give him instruction, who are not professed teachers and who never had a single disciple in that branch of knowledge which he wishes him to acquire-would not such conduct be the height of folly?

Any. Yes, by Zeus, and of ignorance too.

Soc. Very good. And now you are in a position to advise with me about my friend Meno. He has been telling me, Anytus, that he desires to attain that kind of wisdom and-virtue by which men order the state or the house, and honour their parents, and know when to receive and when to send away citizens and strangers, as a good man should. Now, to whom should he go in order that he may learn this virtue? Does not the previous argument imply clearly that we should send him to those who profess and avouch that they are the common teachers of all Hellas, and are ready to impart instruction to any one who likes, at a fixed price? Any. Whom do you mean, Socrates?

9.4.12 FALSE BELIEF

"SOCRATES: Then how on earth, Anytus, could you know what's good and what's worthless about the enterprise, when you have no experience of it at all?"

Plato, "Meno"
Robin Waterfield, "Plato: Meno and Other Dialogues"

During the discussion about whether the mental quality of excellence can be taught, Socrates asserts that it could be the 'Sophists' that teach excellence for pay. The Sophists were teachers of philosophy and rhetoric in ancient Greece. To Plato the Sophists' aim was victory in discourse while Socrates' aim was to generate knowledge. As a result, Sophists often reasoned with clever, but fallacious arguments to win in discourse.

Anytus decries the selection of the Sophists for the teachers of excellence. When Socrates asks Anytus why he has such a poor opinion of the Sophists,

Anytus admits he has no direct experience or knowledge of their ways. In other words, he admits that he doesn't know about them. But Anytus doesn't know that he doesn't really know about the Sophists. This is a false belief – a person holding an opinion they believe to be true of something that is false. Plato then refers to Anytus being "a diviner" foreshadowing the conclusion of the dialectic. Plato is being ironical as he clearly does not think Anytus' reasoning is inspired by the gods.

Plato has used the discussion of whether excellence can be taught to present the pattern of false belief. Later he will succinctly resolve whether excellence can be taught. Now he will assess whether excellence can be learned through habituation.

Soc. You surely know, do you not, Anytus, that these are the people whom mankind call Sophists?

Any. By Heracles, Socrates, forbear! I only hope that no friend or kinsman or acquaintance of mine, whether citizen or stranger, will ever be so mad as to allow himself to be corrupted by them; for they are a manifest pest and corrupting influences to those who have to do with them.

Soc. What, Anytus? Of all the people who profess that they know how to do men good, do you mean to say that these are the only ones who not only do them no good, but positively corrupt those who are entrusted to them, and in return for this disservice have the face to demand money? Indeed, I cannot believe you; for I know of a single man, Protagoras, who made more out of his craft than the illustrious Pheidias, who created such noble works, or any ten other statuaries. How could that A mender of old shoes, or patcher up of clothes, who made the shoes or clothes worse than he received them, could not have remained thirty days undetected, and would very soon have starved; whereas during more than forty years, Protagoras was corrupting all Hellas, and sending his disciples from him worse than he received them, and he was never found out. For, if I am not mistaken,-he was about seventy years old at his death, forty of which were spent in the practice of his profession; and during all that time he had a good reputation, which to this day he retains: and not only Protagoras, but many others are well spoken of; some who lived before him, and others who are still living. Now, when you say that they deceived and corrupted the youth, are they to be supposed to have corrupted them consciously or unconsciously? Can those who were deemed by many to be the wisest men of Hellas have been out of their minds?

Any. Out of their minds! No, Socrates; the young men who gave their money to them, were out of their minds, and their relations and guardians who entrusted their youth to the care of these men were still more out of their minds, and most of all, the cities who allowed them to come in, and did not drive them out, citizen and stranger alike.

Soc. Has any of the Sophists wronged you, Anytus? What makes you so angry with them?

Any. No, indeed, neither I nor any of my belongings has ever had, nor would I suffer them to have, anything to do with them.

Soc. Then you are entirely unacquainted with them?

Any. And I have no wish to be acquainted.

Soc. Then, my dear friend, how can you know whether a thing is good or bad of which you are wholly ignorant?

Any. Quite well; I am sure that I know what manner of men these are, whether I am acquainted with them or not.

Soc. You must be a diviner, Anytus, for I really cannot make out, judging from your own words, how, if you are not acquainted with them, you know about them. But I am not enquiring of you who are the teachers who will corrupt Meno (let them be, if you please, the Sophists); I only ask you to tell him who there is in this great city who will teach him how to become eminent in the virtues which I was just, now describing. He is the friend of your family, and you will oblige him.

Any. Why do you not tell him yourself?

9.4.13 IS EXCELLENT TRANSMITTED BY HABITUATION?

"<u>SOCRATES:</u> As part of this enquiry we're also asking whether the good [excellent] men of present generations or past generations know how to transmit their own excellence to anyone else, or whether excellence is not the kind of thing that can be transmitted to someone or received by anyone else."

Plato, "Meno"
Robin Waterfield, "Plato: Meno and Other Dialogues"

In this section Plato has the characters discuss how the mental quality of excellence could be transmitted to a person through "habituation". The key example is in the case of Themistocles and his son Cleophantus. Themistocles is a legendary statesman and military commander in Greek history. He saved the Athenians and all Greece from the second Persian invasion. But even the great Themistocles could not pass on his excellence to his son Cleophantus through habituation. In Robin Waterfield's *Plato: Meno and Other Dialogues* Socrates quotes the poet Theogenes as saying:

> **_"Never would a bad son have sprung_**
>
> **_from a good sire, for he would have_**
>
> **_heard the voice of instruction; but_**
>
> **_not by teaching will you ever make a_**
>
> **_bad man into a good one."_**

Theogenes succinctly states the reality of the inability of Themistocles to pass on his mental quality of excellence. It is also clear that the mental quality of excellence isn't necessarily imputed through example. The implication is that if even the great Themistocles can't do through habituation, it then it can't be done.

Soc. You surely know, do you not, Anytus, that these are the people whom mankind call Sophists?

Any. By Heracles, Socrates, forbear! I only hope that no friend or kinsman or acquaintance of mine, whether citizen or stranger, will ever be so mad as to allow himself to be corrupted by them; for they are a manifest pest and corrupting influences to those who have to do with them.

Soc. What, Anytus? Of all the people who profess that they know how to do men good, do you mean to say that these are the only ones who not only do them no good, but positively corrupt those who are entrusted to them, and in return for this disservice have the face to demand money? Indeed, I cannot believe you; for I know of a single man, Protagoras, who made more out of his craft than the illustrious Pheidias, who created such noble works, or any ten other statuaries. How could that A mender of old shoes, or patcher up of clothes, who made the shoes or clothes worse than he received them, could not have remained thirty days undetected, and would very soon have starved; whereas during more than forty years, Protagoras was corrupting all Hellas, and sending his disciples from him worse than he received them, and he was never found out. For, if I am not mistaken,-he was about seventy years old at his death, forty of which were spent in the practice of his profession; and during all that time he had a good reputation, which to this day he retains: and not only Protagoras, but many others are well spoken of; some who lived before him, and others who are still living. Now, when you say that they deceived and corrupted the youth, are they to be supposed to have corrupted them consciously or unconsciously? Can those who were deemed by many to be the wisest men of Hellas have been out of their minds?

Any. Out of their minds! No, Socrates; the young men who gave their money to them, were out of their minds, and their relations and guardians who entrusted their youth to the care of these men were still more out of their minds, and most of all, the cities who allowed them to come in, and did not drive them out, citizen and stranger alike.

Soc. Has any of the Sophists wronged you, Anytus? What makes you so angry with them?

Any. No, indeed, neither I nor any of my belongings has ever had, nor would I suffer them to have, anything to do with them.

Soc. Then you are entirely unacquainted with them?

Any. And I have no wish to be acquainted.

Soc. Then, my dear friend, how can you know whether a thing is good or bad of which you are wholly ignorant?

Any. Quite well; I am sure that I know what manner of men these are, whether I am acquainted with them or not.

Soc. You must be a diviner, Anytus, for I really cannot make out, judging from your own words, how, if you are not acquainted with them, you know about them. But I am not enquiring of you who are the teachers who will corrupt Meno (let them be, if you please, the Sophists); I only ask you to tell him who there is in this great city who will teach him how to become eminent in the virtues which I was just, now describing. He is the friend of your family, and you will oblige him.

Any. Why do you not tell him yourself?

Soc. I have told him whom I supposed to be the teachers of these things; but I learn from you that I am utterly at fault, and I dare say that you are right. And now I wish that you, on your part, would tell me to whom among the Athenians he should go. Whom would you name?

Any. Why single out individuals? Any Athenian gentleman, taken at random, if he will mind him, will do far more, good to him than the Sophists.

Soc. And did those gentlemen grow of themselves; and without having been taught by any one, were they nevertheless able to teach others that which they had never learned themselves?

Any. I imagine that they learned of the previous generation of gentlemen. Have there not been many good men in this city?

Soc. Yes, certainly, Anytus; and many good statesmen also there always have been and there are still, in the city of Athens. But the question is whether they were also good teachers of their own virtue;-not whether there are, or have been, good men in this part of the world, but whether virtue can be taught, is the question which we have been discussing. Now, do we mean to say that the good men our own and of other times knew how to impart to others that virtue which they had themselves; or is virtue a thing incapable of being communicated or imparted by one man to another? That is the question which I and Meno have been arguing. Look at the matter in your own way: Would you not admit that Themistocles was a good man?

Any. Certainly; no man better.

Soc. And must not he then have been a good teacher, if any man ever was a good teacher, of his own virtue?

Any. Yes certainly,-if he wanted to be so.

Soc. But would he not have wanted? He would, at any rate, have desired to make his own son a good man and a gentleman; he could not have been jealous of him, or have intentionally abstained from imparting to him his own virtue. Did you never hear that he made his son Cleophantus a famous horseman; and had him taught to stand upright on horseback and hurl a javelin, and to do many other marvellous things; and in anything which could be learned from a master he was well trained? Have you not heard from our elders of him?

Any. I have. Soc. Then no one could say that his son showed any want of capacity?

Any. Very likely not.

Soc. But did any one, old or young, ever say in your hearing that Cleophantus, son of Themistocles, was a wise or good man, as his father was?

Any. I have certainly never heard any one say so.

Soc. And if virtue could have been taught, would his father Themistocles have sought to train him in these minor accomplishments, and allowed him who, as you must remember, was his own son, to be no better than his neighbours in those qualities in which he himself excelled? Any. Indeed, indeed, I think not.

Soc. Here was a teacher of virtue whom you admit to be among the best men of the past. Let us take another,-Aristides, the son of Lysimachus: would you not acknowledge that he was a good man?

Any. To be sure I should.

Soc. And did not he train his son Lysimachus better than any other Athenian in all that could be done for him by the help of masters? But what has been the result? Is he a bit better than any other mortal? He is an acquaintance of yours, and you see what he is like. There is Pericles, again, magnificent in his wisdom; and he, as you are aware, had two sons, Paralus and Xanthippus.

Any. I know.

Soc. And you know, also, that he taught them to be unrivalled horsemen, and had them trained in music and gymnastics and all sorts of arts-in these respects they were on a level with the best-and had he no wish to make good men of them? Nay, he must have wished it. But virtue, as I suspect, could not be taught. And that you may not suppose the incompetent teachers to be only the meaner sort of Athenians and few in number, remember again that Thucydides had two sons, Melesias and Stephanus, whom, besides giving them a good education in other things, he trained in wrestling, and they were the best wrestlers in Athens: one of them he committed to the care of Xanthias, and the other of Eudorus, who had the reputation of being the most celebrated wrestlers of that day. Do you remember them?

Any. I have heard of them.

Soc. Now, can there be a doubt that Thucydides, whose children were taught things for which he had to spend money, would have taught them to be good men, which would have cost him nothing, if virtue could have been taught? Will you reply that he was a mean man, and had not many friends among the Athenians and allies? Nay, but he was of a great family, and a man of influence at Athens and in all Hellas, and, if virtue could have been taught, he would have found out some Athenian or foreigner who would have made good men of his sons, if he could not himself spare the time from cares of state. Once more, I suspect, friend Anytus, that virtue is not a thing which can be taught?

Any. Socrates, I think that you are too ready to speak evil of men: and, if you will take my advice, I would recommend you to be careful. Perhaps there is no city in which it is not easier to do men harm than to do them good, and this is certainly the case at Athens, as I believe that you know.

Soc. O Meno, think that Anytus is in a rage. And he may well be in a rage, for he thinks, in the first place, that I am defaming these gentlemen; and in the second place, he is of opinion that he is one of them himself. But some day he will know what is the meaning of defamation, and if he ever does, he will forgive me. Meanwhile I will return to you, Meno; for I suppose that there are gentlemen in your region too?

Men. Certainly there are.

Soc. And are they willing to teach the young? and do they profess to be teachers? and do they agree that virtue is taught?

Men. No indeed, Socrates, they are anything but agreed; you may hear them saying at one time that virtue can be taught, and then again the reverse.

Soc. Can we call those teachers who do not acknowledge the possibility of their own vocation?

Men. I think not, Socrates.

Soc. And what do you think of these Sophists, who are the only professors? Do they seem to you to be teachers of virtue?

Men. I often wonder, Socrates, that Gorgias is never heard promising to teach virtue: and when he hears others promising he only laughs at them; but he thinks that men should be taught to speak. Soc. Then do you not think that the Sophists are teachers?

Men. I cannot tell you, Socrates; like the rest of the world, I am in doubt, and sometimes I think that they are teachers and sometimes not.

Soc. And are you aware that not you only and other politicians have doubts whether virtue can be taught or not, but that Theognis the poet says the very same thing? Men. Where does he say so?

Soc. In these elegiac verses: Eat and drink and sit with the mighty, and make yourself agreeable to them; for from the good you will learn what is good, but if you mix with the bad you will lose the intelligence which you already have. Do you observe that here he seems to imply that virtue can be taught?

Men. Clearly.

Soc. But in some other verses he shifts about and says: If understanding could be created and put into a man, then they [who were able to perform this feat] would have obtained great rewards. And again:- Never would a bad son have sprung from a good sire, for he would have heard the voice of instruction; but not by teaching will you ever make a bad man into a good one. And this, as you may remark, is a contradiction of the other.

Men. Clearly.

9.4.14 NO TEACHERS MEANS NO STUDENTS

"_SOCRATES: But didn't we agree that any matter for which there are no teachers and no students is not in fact teachable?"_

Plato, "Meno"
Robin Waterfield, "Plato: Meno and Other Dialogues"

In this section, Plato is ending the logically contradicting discussion of whether excellence can be taught. People are born with the ability to pattern-match using their imagination. Therefore, it can't be taught. How could you teach someone something they already can do? Logically, if something can't be taught then no one can be a student of it.

In addition, good guidance is not wholly dependent on knowledge – it is possible without knowledge. This means that excellence isn't knowledge either. The concepts of learning, teaching, student, habituation, natural endowment, and knowledge are now eliminated from the enquiry into excellence. So, the mental quality of excellence must be acquired in some other way.

Soc. And is there anything else of which the professors are affirmed not only not to be teachers of others, but to be ignorant themselves, and bad at the knowledge of that which they are professing to teach? or is there anything about which even the acknowledged "gentlemen" are sometimes saying that "this thing can be taught," and sometimes the opposite? Can you say that they are teachers in any true sense whose ideas are in such confusion?

Men. I should say, certainly not.

Soc. But if neither the Sophists nor the gentlemen are teachers, clearly there can be no other teachers?

Men. No.

Soc. And if there are no teachers, neither are there disciples?

Men. Agreed.

Soc. And we have admitted that a thing cannot be taught of which there are neither teachers nor disciples?

Men. We have.

Soc. And there are no teachers of virtue to be found anywhere?

Men. There are not.

Soc. And if there are no teachers, neither are there scholars?

Men. That, I think, is true.

Soc. Then virtue cannot be taught?

Men. Not if we are right in our view. But I cannot believe, Socrates, that there are no good men: And if there are, how did they come into existence?

Soc. I am afraid, Meno, that you and I are not good for much, and that Gorgias has been as poor an educator of you as Prodicus has been of me. Certainly we shall have to look to ourselves, and try to find some one who will help in some way or other to improve us. This I say, because I observe that in the previous discussion none of us remarked that right and good action is possible to man under other guidance than that of knowledge (episteme);-and indeed if this be denied, there is no seeing how there can be any good men at all.

Men. How do you mean, Socrates?

Soc. I mean that good men are necessarily useful or profitable. Were we not right in admitting this? It must be so.

Men. Yes.

Soc. And in supposing that they will be useful only if they are true guides to us of action-there we were also right?

Men. Yes.

Soc. But when we said that a man cannot be a good guide unless he have knowledge (phrhonesis), this we were wrong.

Men. What do you mean by the word "right"?

9.4.15 TRUE BELIEF

"SOCRATES: True belief, then, is just as good a guide as knowledge, when it comes to guaranteeing correctness of action. This is what we were overlooking before, during our enquiry into the nature of excellence, when we were saying that knowledge is the only good guide of our actions. In fact, though, there's true belief as well."

Plato, "Meno"
Robin Waterfield, "Plato: Meno and Other Dialogues"

In this section, Plato addresses how good guidance is possible without knowledge – through true belief. True belief is when a person forms an opinion that is unjustified by explicit knowledge, but correct, nonetheless. A person can guide good action successfully using either knowledge or true belief. Either will produce the same result in the same given situation. Meno attempts to poke a hole in this logic by stating true belief won't always be right. But Socrates rebuts this notion by saying:

"SOCRATES: What do you mean? Won't someone with true belief always be right, as long as his beliefs are true?"

What Socrates says is true. But Meno is thinking probabilistically about the matter. It isn't highly probable that a person, even such as Themistocles, will have true belief in every situation in the absence of knowledge. In this, Meno makes a very important point. He asks Socrates to explain the higher value placed on knowledge over true belief if they produce the same result.

Soc. I will explain. If a man knew the way to Larisa, or anywhere else, and went to the place and led others thither, would he not be a right and good guide?

Men. Certainly.

Soc. And a person who had a right opinion about the way, but had never been and did not know, might be a good guide also, might he not?

Men. Certainly.

Soc. And while he has true opinion about that which the other knows, he will be just as good a guide if he thinks the truth, as he who knows the truth?

Men. Exactly.

Soc. Then true opinion is as good a guide to correct action as knowledge; and that was the point which we omitted in our speculation about the nature of virtue, when we said that knowledge only is the guide of right action; whereas there is also right opinion.

Men. True.

Soc. Then right opinion is not less useful than knowledge?

Men. The difference, Socrates, is only that he who has knowledge will always be right; but he who has right opinion will sometimes be right, and sometimes not.

Soc. What do you mean? Can he be wrong who has right opinion, so long as he has right opinion?

Men. I admit the cogency of your argument, and therefore, Socrates, I wonder that knowledge should be preferred to right opinion-or why they should ever differ.

9.4.16   EXPANSION OF KNOWLEDGE

"*SOCRATES:* *When true beliefs are anchored, they become pieces of knowledge and they become stable. That's why knowledge is more valuable than true belief, and the difference between the two is that knowledge has been anchored.*"

Plato, "Meno"
Robin Waterfield, "Plato: Meno and Other Dialogues"

In this section, Plato explains the higher value of knowledge or scientific information than that of true belief. True belief is an accurate perception of something without explicit knowledge. The problem is that this true belief may not continue to apply to changing conditions. It is only applicable to the specific circumstance and situation in which it was formed. The possessor of the true belief does not know the underlying concept(s) that would enable its application in any situation or circumstance.

However, true belief is an understanding of a pattern in nature at a given point in time. If the true belief is further refined by reason, one can work out the logic chain to connect it to existing knowledge. This makes this true belief into knowledge.

What Plato is describing is the pattern of excellence. This is the pattern by which homo sapiens pattern-match using imagination to create true belief. Homo sapiens then use reason to build the logic chain to existing knowledge. This transforms true belief into knowledge – a more valuable form of information. This enables homo sapiens to adapt as circumstances change. This excellence enables homo sapiens to perfect their fitness in any ecosystem found in nature.

Soc. And shall I explain this wonder to you?

Men. Do tell me.

Soc. You would not wonder if you had ever observed the images of Daedalus; but perhaps you have not got them in your country?

Men. What have they to do with the question?

Soc. Because they require to be fastened in order to keep them, and if they are not fastened they will play truant and run away.

Men. Well. what of that?

Soc. I mean to say that they are not very valuable possessions if they are at liberty, for they will walk off like runaway slaves; but when fastened, they are of great value, for they are really beautiful works of art. Now this is an illustration of the nature of true opinions: while they abide with us they are beautiful and fruitful, but they run away out of the human soul, and do not remain long, and therefore they are not of much value until they are fastened by the tie of the cause; and this fastening of them, friend Meno, is recollection, as you and I have agreed to call it. But when they are bound, in the first place, they have the nature of knowledge; and, in the second place, they are abiding. And this is why knowledge is more honourable and excellent than true opinion, because fastened by a chain.

Men. What you are saying, Socrates, seems to be very like the truth.

Soc. I too speak rather in ignorance; I only conjecture. And yet that knowledge differs from true opinion is no matter of conjecture with me. There are not many things which I profess to know, but this is most certainly one of them.

Men. Yes, Socrates; and you are quite right in saying so.

Soc. And am I not also right in saying that true opinion leading the way perfects action quite as well as knowledge?

Men. There again, Socrates, I think you are right.

Soc. Then right opinion is not a whit inferior to knowledge, or less useful in action; nor is the man who has right opinion inferior to him who has knowledge?

Men. True.

9.4.17 GIFT OF THE GODS

"SOCRATES:…When they raise important issues in their speeches and see them through to a successful conclusion, despite not understanding anything they say, they're inspired: they've been taken over by the gods."

Plato, "Meno"
Robin Waterfield, "Plato: Meno and Other Dialogues"

In this section, Plato has the characters Socrates and Meno summarize the logic system developed throughout the dialectic. Either only true belief or knowledge enable humans to generate good guidance for their city state. But neither true belief nor knowledge is a natural endowment. Therefore, excellence cannot be a natural endowment.

Since excellence is not a natural endowment, it cannot be any special intellectual capacity. Also, excellence is not knowledge, so it cannot be taught. Therefore, there are no teachers and no students of excellence. Ultimately, this is why excellence cannot be acquired by teaching, habituation, or natural inheritance.

Through the process of elimination Plato has determined what excellence is not. This leaves only one remaining possibility for what excellence is – true belief. But true belief can only be produced through imagination. It is the imagination of politicians such as Themistocles that enables them to offer good guidance and lead city states well. Socrates asserts that since politicians achieve excellence without knowledge it must be divinely inspired. He compares the true belief of politicians to that of soothsayers, oracles, and prophets. In Robin Waterfield's *Plato: Meno and Other Dialogues,* Plato has Socrates explicitly state this conclusion:

"SOCRATES: So, Meno, our argument has led us to suppose that the excellence [imagination] of good people comes to them as a dispensation awarded by the gods."

The ability to generate true belief is a gift of the gods and that people deem such politicians to be godlike. If any politician could teach others to develop their own true beliefs it would be the act of a god. This makes those alone wise who can teach others to develop true belief by using their imagination. Other men, such as Sophists, would be only shadows by comparison.

Soc. And surely the good man has been acknowledged by us to be useful?

Men. Yes.

Soc. Seeing then that men become good and useful to states, not only because they have knowledge, but because they have right opinion, and that neither knowledge nor right opinion is given to man by nature or acquired by him -- do you imagine either of them to be given by nature?

Men. Not I.

Soc. Then if they are not given by nature, neither are the good by nature good?

Men. Certainly not.

Soc. And nature being excluded, then came the question whether virtue is acquired by teaching?

Men. Yes.

Soc. If virtue was wisdom [or knowledge], then, as we thought, it was taught?

Men. Yes.

Soc. And if it was taught it was wisdom?

Men. Certainly.

Soc. And if there were teachers, it might be taught; and if there were no teachers, not?

Men. True.

Soc. But surely we acknowledged that there were no teachers of virtue?

Men. Yes.

Soc. Then we acknowledged that it was not taught, and was not wisdom?

Men. Certainly.

Soc. And yet we admitted that it was a good?

Men. Yes.

Soc. And the right guide is useful and good?

Men. Certainly.

Soc. And the only right guides are knowledge and true opinion-these are the guides of man; for things which happen by chance are not under the guidance of man: but the guides of man are true opinion and knowledge.

Men. I think so too.

Soc. But if virtue is not taught, neither is virtue knowledge.

Men. Clearly not.

Soc. Then of two good and useful things, one, which is knowledge, has been set aside, and cannot be supposed to be our guide in political life.

Men. I think not.

Soc. And therefore not by any wisdom, and not because they were wise, did Themistocles and those others of whom Anytus spoke govern states. This was the reason why they were unable to make others like themselves-because their virtue was not grounded on knowledge.

Men. That is probably true, Socrates.

Soc. But if not by knowledge, the only alternative which remains is that statesmen must have guided states by right opinion, which is in politics what divination is in religion; for diviners and also prophets say many things truly, but they know not what they say.

Men. So I believe.

Soc. And may we not, Meno, truly call those men "divine" who, having no understanding, yet succeed in many a grand deed and word?

Men. Certainly.

Soc. Then we shall also be right in calling divine those whom we were just now speaking of as diviners and prophets, including the whole tribe of poets. Yes, and statesmen above all may be said to be divine and illumined, being inspired and possessed of God, in which condition they say many grand things, not knowing what they say.

Men. Yes.

Soc. And the women too, Meno, call good men divine-do they not? and the Spartans, when they praise a good man, say "that he is a divine man."

Men. And I think, Socrates, that they are right; although very likely our friend Anytus may take offence at the word.

Soc. I do not care; as for Anytus, there will be another opportunity of talking with him. To sum up our enquiry-the result seems to be, if we are at all right in our view, that virtue is neither natural nor acquired, but an instinct given by God to the virtuous. Nor is the instinct accompanied by reason, unless there may be supposed to be among statesmen some one who is capable of educating statesmen. And if there be such an one, he may be said to be among the living what Homer says that Tiresias was among the dead, "he alone has understanding; but the rest are flitting shades"; and he and his virtue in like manner will be a reality among shadows.

Men. That is excellent, Socrates.

Soc. Then, Meno, the conclusion is that virtue comes to the virtuous by the gift of God. But we shall never know the certain truth until, before asking how virtue is given, we enquire into the actual nature of virtue. I fear that I must go away, but do you, now that you are persuaded yourself, persuade our friend Anytus. And do not let him be so exasperated; if you can conciliate him, you will have done good service to the Athenian people.

9.5   CONCLUSION

In the *Meno* Plato describes the process by which homo sapiens use imagination to artificially adapt to their environment. It is through this process that our species has become evolutionarily fitted to so many ecosystems on the planet. This process ultimately leads to our species perfecting their artificial adaptations in any ecosystem. We can see this in the distinct adaptations of Alaskan tribes, Amazonian tribes, the Iroquois tribes, German tribes, Australian tribes, Chinese tribes, and the first African tribes.

The species homo sapiens has artificially adapted to survive and thrive wherever possible. This is why we are the apex species on planet earth. And then we adapted to transcend those environments to create the modern world. Now rather than adapting ourselves to the environment we are adapting the environment to ourselves. In *Meno* Plato is describing humanity's most important evolutionary advantage. What else could possibly be homo sapiens excellence but imagination?

In Plato's mind the power of developing true belief can only be a gift from the gods. But remember, the ancient Greeks believed their gods were found in nature. To Plato, the mental quality that allows homo sapiens to acquire excellence is found in nature – evolution.

So, Plato's conclusion is correct after all as homo sapiens imagination is a gift of nature or the gods. It is the fire Prometheus gave to man. But of course, he's correct – he's Plato. Plato just didn't have the more precise scientific knowledge of genetics and neurology we do today.

Plato is the man that documented the ancient Greek true belief about imagination in *Meno*. Now finally, we can chain the true beliefs expressed in the *Meno* to our existing pattern of homo sapiens knowledge. So, we can transform the true beliefs of the ancient Greeks into scientific knowledge. We achieved this result simply by using our combining existing knowledge with our imagination. Somewhere, Plato is smiling at us – we recollected something lost to history using only our imagination without having to be taught. The pattern is now matched – evolutionary theory and ancient Greek philosophy. Our cipher of Darwin's book *On the Origin of Species* has proven a valid one to decode ancient Greek philosophy.

10

CHAPTER 10: DECODING ANCIENT GREEK PHILOSOPHY

"For better or worse, Socrates, Plato, and Aristotle engineered the Western mind. Above all, they formed part of a movement that stood at the crossroads of mythological and scientific-rational thought, at the crossroads of mythos and logos."

Neel Burton, "The Gang of Three"

10.1   INTRODUCTION

"Linnaeus and Cuvier have been my two gods, though in very different ways, but they were mere schoolboys to old Aristotle."

Charles Darwin, "The Life & Letters of Charles Darwin"

Ancient Greek philosophy is now decoded. Charles Darwin himself gave us the clue. He compared the naturalists of his time to Aristotle. Notice how Charles Darwin equates Linnaeus and Cuvier to gods. That is an expression of his ancient Greek mindset.

We will now "recollect" or search for that pattern in the ancient Greek's "writings on nature". Remember, when the ancient Greeks say "nature" they really mean the "world". They included everything in the "world" – natural and artificial. To the ancient Greeks to unconceal the pattern of the gods and act in harmony with that pattern was to be godlike. This was the aspiration for both an individual and a city state. They sought to achieve a kind of divine perfection in homo sapiens affairs – and for the ancient Greeks the divine was found in nature.

10.2   ON THE ORDER OF NATURE – PARMENIDES

10.2.1 INTRODUCTION

"In stark contrast to Heraclitus, who held that everything is in a state of flux, Parmenides of Elea (c. 515 – c. 440 BCE) argued that nothing ever changes."

Neel Burton, "The Gang of Three"

Parmenides of Elea was the first known to discover the pattern of the gods in nature. He called it the "One" pattern. This one pattern was the aggregation of many patterns in both physics and evolution. Imagine Parmenides sitting on his porch watching the sun, the seasons, and all species following cyclical patterns. He must have concluded that a set of laws governed the natural order of the world.

To understand "The One" pattern is not complicated. You likely have already seen it in the original Jurassic Park movie. The underlying pattern of the natural ecosystem on Jurassic Park's Island was identical to that of modern-day Africa and North America during the ice age. You naturally pattern-match the dynamics of all three without realizing it. This is because homo sapiens adapted to survive in that pure natural ecosystem pattern long ago.

The "One" pattern emerged again and again during earth's history. Together the laws of physics and the process of evolution produce this result. It is as inevitable as the sun rising each day. As we stated earlier – "**it is the only game in town**." Parmenides perceived this one pattern with only his imagination – amazing.

10.2.2 NATURAL LAW AND UNIVERSAL ORDER

"Rejoice, for certainly no adverse Moira [fate] has urged you to travel this way – which is far from the path travelled by men – but [you have been sent] by Θέμις [divine law] and Δίκη [universal order]."

Parmenides, "On the Order of Nature" [Asram Vidya]

This passage refers to the laws of physics (divine law) which create the order of the universe. Part of this universal order is the process of evolution in nature. Specifically, Parmenides is referring to Dike (Δίκη) the ancient Greek goddess of justice. She creates order in human affairs based on law. This is meaningful in the context within which Parmenides views this subject. We will refer to this point later.

10.2.3 THE ONE PATTERN OF NATURE

"[From this is follows that] there remains only one discourse of the Way (ὁδός) that is...that Being is unborn, (ἀνώλεθρόν) uncorruptible; indeed it is whole (οὖλον) in its entirety, motionless, and endless (ἀτέλεστον). It never was and will never be because it is now all together one (ἕν and therefore not many), and continuous (συνεχές).

Parmenides, "On the Order of Nature" [Asram Vidya]

In this passage Parmenides directly describes the "One" pattern. It is the repetitive pattern created by the law of physics and the process of evolution. This pattern has never physically existed. It can only be perceived in the mind's eye.

However, as we discussed earlier, variations of this same pattern have reoccurred on earth. It can never be corrupted while the dynamics of the planet remain stable. The word "Being" has been mistranslated in the text in certain instances. The text likely alternately describes "Being" and "Patterns". But the translator did not successfully translate this meaning to us.

10.2.4   THE PATTERNS OF NATURE

"There, indeed, was I taken, there the mares conducted me, drawing my chariot, while maidens pointed out the way ὁδός [idea that underlies the form]."

Parmenides, "On the Order of Nature" [Asram Vidya]

Parmenides accurately perceived that the "One" pattern had many sub patterns (e.g., evolution, etc.). For example, each species in nature has a visible pattern called a phenotype. This is how we can classify many similar organisms into a single species. Another sub pattern is the seasonal weather of the planet. Farmers structure their own patterns of behavior around these seasonal weather patterns.

Parmenides understood that today's patterns will eventually change. They have many times before. So, you must look past these variations to see the underlying concept or pattern. Armed with this knowledge you can discern the specific patterns of the planet. For these patterns will be repeated wherever you live.

Let's revisit the Jurassic Park example. The concepts or patterns of "carnivore" and "herbivore" repeat across earth's history. Tyrannosaurus rex and lions are the carnivores. Triceratops and rhinoceros are herbivores which are also eerily similar in phenotype. To Parmenides to understand these underlying patterns in nature is the "Way of Truth". Whose truth? The truth of the gods. The truths they jealously guard from mere mortal men.

10.2.5   IMAGINE THE PATTERNS

"Observe with noetic intuition how things which are absent (ἀπεόντα) are in truth equally present: in fact, you will not be able to separate Being [τὸ ἐὸν] from Being (ἐόντος) either in a complete [cosmic] dissolution or when it condenses."

Parmenides, "On the Order of Nature" [Asram Vidya]

In this passage Parmenides discusses how the patterns in nature can be unconcealed. By "noetic intuition" he means your imagination. Through your imagination you can unconceal the patterns that are eternally present in nature. You can do this by mentally separating the underlying conceptual pattern from the visible variation in nature. To the ancient Greeks that is the true power of homo sapiens' imagination.

This was the true offense committed by Prometheus when he gave man fire. With the gift of fire man developed imagination – and with that came the ability to discern the secrets of the gods in nature. This is what wisdom truly means. It is thought that Prometheus' very name is synonymous with imagination as meaning "forethought". The definition of forethought is "thinking or planning out in advance." This is only possible with imagination.

This is why nemesis, the goddess of retribution for homo sapiens hubris or arrogance before the gods, is mentioned when Parmenides is at the "Gate of the Ways of Day and Night". For the Olympian gods overthrew the Titans to gain supremacy. With the gift of imagination that can unconceal the secrets of the gods, homo sapiens became a threat to the gods' supremacy.

It is a foreign idea to western thinking to challenge God for that is what Satan did. But for the ancient Greeks becoming godlike was the ultimate form of excellence. The mighty Hercules was in the end deified and took his place on Mount Olympus among the gods. Other mortals, such as Hercules and Psyche, also became gods. To be deified for your excellence was the ultimate honor and glory for the ancient Greeks.

10.2.6 PATTERN-MATCHING

"The same is true of thinking and of that on account of which [οὕνεκεν in the text] there is thought, since without Being [patterns], by which it is made manifest, you will not find thought...For it, all of those things that the mortals decided, convinced that they were true [that is absolute], will be names: being born and vanishing (ὄλλυσϑαι), being and non-being, the changing position and the variation of luminous colour [images]."

Parmenides, "On the Order of Nature" [Asram Vidya]

In this passage Parmenides describes the process of thought as based on pattern-matching. We already covered this in the *Meno*. Basically, Parmenides says your thoughts are based on patterns so without patterns thought is not practically possible. He is generally correct as concepts (patterns) are the building blocks of thought. We will discuss later how Plato built on this concept in his various philosophical theories.

10.2.7 THE WAY OF TRUTH

"The decision is therefore inevitable: one of the ways must be abandoned, for it cannot be travelled or expressed, being the way of untruth, while the other way is, and is therefore true [this is the ὁδός which leads to the ἐὸν]."

Parmenides, "On the Order of Nature" [Asram Vidya]

To Parmenides the only truth which there can ever be is the "One" incorruptible pattern. All else is only opinion or is inaccurate. Until you perceive the world through this lens of truth, you are only dimly seeing shadows on the wall of the real object. This is equivalent to Plato's "Allegory of the Cave". It is what enables homo sapiens to fully understand nature.

We call this understanding scientific knowledge, such as the law of gravity and the theory of evolution. This is what knowledge means in ancient Greek – scientific understating of nature. These items of scientific knowledge are pieces of the pattern Parmenides is described. Until you can fully unconceal pieces of the pattern, such as natural evolution, you possess only an opinion of the natural world – not the truth.

For that is what truth means in ancient Greek – aletheia or to unconceal the secrets of the gods (nature). Therefore, the "Way of Truth" is to walk in the footsteps of the gods who created the natural world. This is to acquire the secrets of how the divine laws shape the natural order in which homo sapiens exist. This is what Parmenides means by "The Way of Truth".

10.3   ON NATURE – HERACLITUS

10.3.1 INTRODUCTION

"Heraclitus' big idea is that everything is in a constant state of flux, as epitomized by his saying that you cannot step twice into the same river. Everything flows, everything moves…the waters are no longer the same, and neither are we."

Neel Burton, "The Gang of Three"

Heraclitus was a relative contemporary of Parmenides. However, as a philosopher, he disagreed with Parmenides. Heraclitus viewed the visible world as constantly changing – and he was correct. The world is constantly changing as nature is inherently dynamic. It is a reality we are all intimately familiar with.

Together the two perspectives of Parmenides and Heraclitus are equivalent to a saying we are all familiar with – "the more things change, the more they stay the same." It seems paradoxical, but it is not since the laws and order of the planet remain constant. The same patterns reemerge in nature again and again. But each variation is different and dynamic or constantly changing. Parmenides and Heraclitus were both correct.

10.3.2   NATURAL LAW AND UNIVERSAL ORDER

"Those who speak intelligence must never stop believing in that which is common to us all; they must do so even more strongly than a city has conviction for its laws, because all human laws are dependent upon one divine law, and divine law rules as far as it desires, and it is abundant and sufficient for everyone."

Heraclitus, "On Nature" [Veronica Ambrose]

In this passage, Heraclitus describes the reality of homo sapiens' existence. The planet is shaped by the laws of physics and the process of evolution. Therefore, we create laws in harmony with natural law. But the ancient Greeks meant this much more literally than we understood to be the case.

Heraclitus is literally saying to base homo sapiens law on divine or natural law. This is exactly what the Greeks attempted to do. Therefore, the design of any system of homo sapiens law would need to be in harmony with natural law – as with evolution.

10.3.3   THE ONE PATTERN OF NATURE

"Listening to universal reason, rather than to my account, it is wise to acknowledge that all things are one."

Heraclitus, "On Nature" [Veronica Ambrose]

In this passage, Heraclitus acknowledges that there is "One" pattern to the world similar to Parmenides' theory. Heraclitus just disputes the eternal and unchanging nature of Parmenides. As Heraclitus states below in Veronica Ambrose's translation of "On Nature":

"Understanding the intelligent force by which all things are governed, through all things, is one true wisdom."

So, Heraclitus agreed with Parmenides that there was one pattern to nature – that to unconceal it is the source of scientific knowledge.

10.3.4 THE DYNAMISM OF NATURE

"A person cannot step into the same river twice because it is not the same river, and they are not the same person."

Heraclitus, "On Nature" [Veronica Ambrose]

Heraclitus understood the dynamic nature of the world. He saw that the rocks on the ground, the fish in the sea, and the birds in the air, and the continents themselves were all constantly changing. Everything on earth will continue to change as long as the laws of physics and the process of evolution persist.

10.3.5 CONCEALED SCIENTIFIC KNOWLEDGE

"Nature loves to conceal itself."

Heraclitus, "On Nature" [Veronica Ambrose]

In this passage Heraclitus alludes to the concept of concealment. The laws of physics and the process of evolution have been concealed from homo sapiens. This is where the term aletheia or unconcealedness derives its meaning in ancient Greek language. But in the Greek cultural context, he means the gods love to conceal their secrets. The secrets they used to create and to govern the world.

10.3.6 THE STRUGGLE FOR EXISTENCE

"Homer was wrong in saying "If only strife would perish from among the gods and human beings." He did not understand that he was hoping for the destruction of the universe, because if his pleases were heard, all things would cease to exist...We must know that war is common to all, and that

strife is just, and that all things come into existence and pass away in strife."

Heraclitus, "On Nature" [Veronica Ambrose]

In this passage Heraclitus describes what Charles Darwin called "the struggle for existence". It is conceptually identical. However, the term strife has a specific meaning in ancient Greek thinking. The ancient Greek goddess Eris' name is translated as "strife". Eris is the Greek goddess of strife and discord. It was the goddess Eris that famously caused the Trojan War by sowing discord between the goddesses Hera, Athena, and Aphrodite.

10.4 METAPHYSICS – ARISTOTLE

10.4.1 INTRODUCTION

"The Metaphysics [meta ta physika, 'after the physics'] is so named for coming after the Physics in the Corpus Aristotelicum. Aristotle himself did not refer to its subject matter, which he defined as 'being qua being', as 'metaphysics' but as 'first philosophy."

Neel Burton, "The Gang of Three"

Metaphysics is Aristotle's discussion of the concepts we have already discussed with one exception. The concepts are being repeated to show their prevalence in ancient Greek philosophical thinking. We will briefly describe each of the previously detailed concepts.

10.4.2 NATURAL LAW AND UNIVERSAL ORDER

"Further, besides sensible things and Forms [patterns] he says there are objects of mathematics, which occupy an intermediate position, differing from sensible things in being eternal and unchangeable, from Forms [patterns] in that there are many alike while the Form [pattern] itself is in each case unique."

Aristotle, "Metaphysics"

In this passage Aristotle describes Plato's conception of the world. Plato had a true belief that mathematics (physics) lies between the visible world and the pure patterns described earlier. Plato is in fact correct. In addition, Aristotle succinctly states the purpose of ancient Greek philosophy below:

"It is right also that philosophy should be called knowledge of truth [unconcealed nature]. For the end of theoretical knowledge is truth [unconcealed nature], while that of practical knowledge is action (for even if they consider how things are, practical men do not study the eternal, but what is relative and in the present)."

Greek philosophy's express purpose was to unconceal the eternal secrets of the gods (nature). The end of this unconcealment was the generation of theoretical knowledge such as the theory of natural evolution. That is exactly what the ancient Greeks were in the process of unconcealing.

10.4.4 THE ONE PATTERN OF NATURE

"Evidently, then, these also consider that number is the principle both as matter for things and as forming both their modifications and their permanent state...and that the One proceeds from both of these...and number from the One; and that the whole heaven, as has been said, is numbers."

Aristotle, "Metaphysics"

In this passage Aristotle restates the true belief of Greek philosophers in the "One" pattern.

10.4.5 THE PATTERNS OF NATURE

"...Plato accepted his teaching, but held that the problem applied not to sensible things but to entities of another kind for this reason, that the common definition could not be a definition of any sensible thing, as they were always changing. Things of this sort, then, he called Ideas, and sensible things, he said, were all named after these, and in virtue of a relation to these; for the many existed by participation in the Ideas that have the same name as they."

Aristotle, "Metaphysics"

In this passage, Aristotle restates the true belief of Greek philosophers in the sub patterns in nature of the "One" pattern".

10.4.6 NATURAL VS ARTIFICIAL PRODUCTS

"Things which are formed in nature are in the same case as these products of art...The natural things which (like the artificial objects previously considered) can be produced spontaneously are those whose matter can be moved by itself in the way in which the seed usually moves it; those things which have not such matter cannot be produced except from the parent animals themselves."

Aristotle, "Metaphysics"

In this passage Aristotle discusses a contrast between the productions of nature and the productions of homo sapiens. Objects produced by nature are termed "natural things". Objects produced by homo sapiens are termed "artificial things." Here we see a logical connection to the concept of artificial selection. Aristotle describes exactly what "artificial objects", or "art" are:

"Now art arises when many notions gained by experience one universal judgement about a class of objects is produced...but to judge that it has done good to all persons of a certain constitution, marked off as one class, when they were ill of this disease...this is a matter of art."

Here, Aristotle describes the medical profession in terms of treating patients. So, "art" to Aristotle isn't just paintings or sculptures as we commonly use the word. Instead the term "art" is being used to describe anything intentionally produced by homo sapiens, not found in nature. The "things" produced by homo sapiens are "artificial" as compared to natural "things".

The ancient Greeks had another name for art with which we are familiar – "techne". Techne means 'art or craft'. In evolutionary terms this approximately translates to structural (art) or behavioral (craft) phenotypic adaptations. But the term art would also apply to paintings, plays, etc. as we mean today.

So, the term art encompasses anything and everything homo sapiens intentionally produces that is also not found in nature. The art is produced using our imagination. So, logically, this must be directly related to the process of artificial selection.

10.5 PHYSICS – ARISTOTLE

10.5.1 INTRODUCTION

"The Physics [ta physika, 'natural things'] is a study of nature, that is, of natural objects that are subject to change, and of change itself, which, for Aristotle, is closely tied to motion.

Neel Burton, "The Gang of Three"

Physics is one of the ancient Greek's "writings on nature". This work of art produced by Aristotle discusses the natural world. Aristotle again discusses the same concepts of; 1) natural law and universal order, 2) the "One" pattern, 3) the patterns in nature. But the primary purpose of philosophical writing is to discover the principles and causes of change (motion) in nature.

10.5.2 THE PATTERN OF EVOLUTION

"Now intelligent action is for the sake of an end; therefore the nature of things also is so. Thus if a house, e.g. had been a thing made by nature, it would have been made in the same way as it is now by art; and if things made by nature were made also by art, they would come to be in the same way as by nature. Each step then in the series is for the sake of the next; and generally art partly completes what nature cannot bring to a finnish, and partly imitates her."

Aristotle, "Physics"

In this passage, Aristotle describes the general pattern of evolution for both natural and artificial productions. Each is adaption becomes perfected to fit the specific end for which it aims. This process of adaption is a series of steps towards that specific end. In nature, that end is survival and reproduction. In most, but not all, cases, it is the same for artificial productions as selected by homo sapiens for use. This is a bedrock concept of evolutionary theory.

Aristotle describes art as completing and imitating nature – but only partly. Think of a winter coat in a snowstorm. Our natural skin is not fit to survive frozen temperatures in a snowstorm – unlike a polar bear's fur. So, we generate artificial production or art in the form of a winter coat.

We have now partially completed the evolutionary adaption of our species to the snowstorm. We have extended our inherited homo sapiens phenotype artificially. We make ourselves evolutionarily fit and perfected to survive winter conditions. We have also imitated the natural adaptation of the polar bear's fur. This is what Aristotle meant by both completing and imitating nature.

10.5.3 NATURAL & ARTIFICIAL SELECTION

"Wherever then all the parts came about just what they would have been if they had come be for an end, such things survived, being organized spontaneously in a fitting way; whereas those which grew otherwise perished and continue to perish, as Empedocles says his 'man-faced ox-progeny' did."

Aristotle, "Physics"

In this passage, Aristotle literally describes the concept of natural selection. This is attributed to the Greek philosopher Empedocles. Empedocles attributes this process to the divine, which we understand means nature. Here you have the concepts of evolution, fitness, survival of the fittest, and the extinction of a species all in one long sentence. Then we combine this with the concept within Aristotle's statement from *Physics* below:

"It is absurd to suppose that purpose is not present because we do not observe the agent deliberating. Art does not deliberate. If the ship-building art were in the wood, it would produce the same results by nature. If, therefore, purpose is present in art, it is present in nature. The illustration is a doctor doctoring himself: nature is like that...It is plain then that nature is a cause, a cause that operates for a purpose."

Here, Aristotle describes a mechanism that operates with a purpose. Note that the mechanism need not be visible to the human eye. He describes the mechanism by which nature's purpose is achieved as equivalent to a "doctor doctoring himself". Combine this concept with the struggle for existence as described by Heraclitus, and you have the missing ingredient. What is strife if not competition? Therefore, competition drives the mechanism of natural selection by which nature acts as a "doctor doctoring himself."

Aristotle also contrasts natural selection with artificial selection. Art does not possess natural instincts that drive action or motion. Homo sapiens must provide the motion to the artificial production. Aristotle believes this is a key distinction between artificial and natural productions.

10.5.4 CONCEPTUAL SELECTION

"The arts, therefore, which govern the matter and have knowledge are two, namely the art which uses the product and the art which directs the production of it...For the helmsman knows and prescribes what sort of form a helm should have, the other from what wood it should be made and by what means of what operations. In the products of art, however, we make the material with a view to the function, whereas in the products of nature the matter is there all along."

Aristotle, "Physics"

In this passage, Aristotle describes a type of selection during the production of art. It is a form of conceptual pattern selection. It is a similar concept to the process of sexual selection – a pattern is selected. That is what breeders do during the process of selective breeding – select a pattern they want to produce by mating two natural organisms. As Aristotle states below in *Physics*:

"So it is with all other artificial products. None of them has in itself the source of its own production."

Selective breeding is the art of shaping the productions of natural organisms – which have the source of their own production. Homo sapiens can only work within the constraints of the genetic information that paired natural organisms inherently possess.

So, selective breeding is not artificial selection. Selective breeding is a combination of the processes of conceptual selection and sexual selection. The result produced is then a new natural organism that is a form of art. It is no different from selectively breeding plants to produce a beautiful aesthetic garden – an art form. In neither case is the process of natural selection enacted to produce the result.

10.7 NICHOMACHEAN ETHICS – ARISTOTLE

10.7.1 INTRODUCTION

"Aristotle's Ethics, which seeks to determine the nature of human happiness [perfection], is closely connected to his Politics, which, in eight more books, seeks to determine the form of government that can maximize happiness [perfection]."

Neel Burton, "The Gang of Three"

The Nichomachean Ethics is Aristotle's work on the application of evolutionary concepts to homo sapiens' behavior. His intent was to perfect homo sapiens' behavior within the context of the ancient Greek city state. Aristotle believed that the intentional development of artificial behavioral adaptations was critical – he said it was the only way to acquire a perfected character. A person's character is comprised of a combination of natural and artificial instincts.

The goal of the ancient Greeks was for homo sapiens to achieve a perfected character. This required homo sapiens to adapt character to fit a specific society. An individual must perfect his or her character to fit personal and professional relationships. If this was achieved, then harmony would result. The Greeks had a word for this – "eudaimonia or godlike". It meant that homo sapiens was in perfect harmony with nature – ultimately with the gods on Mount Olympus themselves.

10.7.2 ARTIFICIAL SELECTION

"And as in the Olympic Games it is not the most beautiful and the strongest that are crowned but those who compete (for it is some of these that are victorious), so those who act win, and rightly win, the noble and good things in life."

Aristotle, "Nichomachean Ethics"

In this passage Aristotle describes an artificial variation of the mechanism of natural selection. The designers of the Olympic Games conceptually selected such a form to maximize competition. It was applied to specific events that did not require a scoring system (e.g., chariot racing, wrestling, etc.).

The Olympic Games and modern-day sports apply natural evolution's "survival of the fittest" pattern. Homo sapiens utilize their combined natural and artificial adaptations to avoid elimination (i.e., extinction) in competitive sports. Only one can survive in the struggle for competitive existence. There can only be one champion.

Only the ecosystem within which competition takes place is artificial – rules and regulations, and the environment (playing area) selected. However, many competitive sports still occur in a natural environment such as a grass field (e.g., football, baseball, rugby, soccer, etc.). But the selection through direct competition is still the same.

10.7.3 ARTIFICIAL FITNESS

"For, as the name itself suggests, it is fitting expenditure involving largeness of scale. But the scale is relative; for the expense of equipping a trireme is not that same as that of heading a sacred embassy. It is what is fitting, then, in relation to the agent, and to the circumstances and the object."

Aristotle, "Nichomachean Ethics"

In this passage Aristotle is describing a variation of natural fitness – artificial fitness. Actions, such as making an expenditure, must be relatively fitted to their ends. He also references two forms of artificial productions for which these expenditures might be made – equipping a trireme and heading a sacred embassy. Aristotle believes that a means must be fitted to achieve its end. Each adaptation is only a means to an end – not an end in itself.

Aristotle has also highlighted an artificial structural adaptation (a trireme) and an artificial behavioral adaptation (heading a sacred embassy). A trireme is an artificial structural adaptation that enables man to cross oceans like a whale. Heading a sacred embassy requires a set of artificial behavioral adaptations necessary to successfully conduct diplomacy. By the same logic, both types of artificial adaptations must be fitted to achieve their specific ends.

This is a core concept of Aristotle's Nichomachean Ethics, and it is a core evolutionary concept. If a species is to reach natural perfection, its adaptations must be optimally fitted to its natural ecosystem. As Aristotle stated in *Nichomachean Ethics*:

"Therefore, if this is true in every case, the virtue [fitness] of man also will be the state of character which makes a man good [perfected] and which makes him do his own work well [godlike]."

Aristotle means to be in harmony with nature is godlike. Only when homo sapiens produce harmony in relationships is peace possible in our affairs.

10.7.4   ARTIFICIAL EQUILIBRIUM

"The just, then, must be both intermediate and equal and relative (i.e. for certain persons). And since the equal intermediate it must be between certain things (which are respectively greater and less); equal, it involves two things; qua just, it is for certain people...Further, this is plain from the fact that awards should be 'according to merit'; for all men agree that what is just in distribution must be according to merit in some sense...The just, then, is a species of the proportionate...."

Aristotle, "Nichomachean Ethics"

The evolutionary concept of equilibrium is central to ancient Greek philosophy. Only by creating and maintaining a relative artificial equilibrium can a city-state avoid strife and discord. It is strife and discord that drives the Polybian governmental cycle; monarchy-tyranny-aristocracy-oligarchy-democracy-ochlocracy (mob rule). Inherent in this cycle is chaos, violence, and injustice. The Greeks had watched this cycle repeat for centuries in city states across Sicily, southern Italy, Greece, and western Turkey.

This is why relative justice is a core concept to ancient Greek philosophy. Homo sapiens in a society need to believe that decisions of judges are fair and equitable. It was hoped that discord between the parties would naturally dissipate. Then harmony, peace, and prosperity could again be possible.

This is critical as the ancient Greeks believed prosperity prevented political unrest. If everyone in the community prospered or had the opportunity to expand, peace would result. This peace would prevent the triggering of the process of the Polybian cycle. This typically led to discord and violence. This is why the Founders of the United States included "pursuit of happiness" as a core goal of their new nation. It is another way of saying evolutionary expansion – in this they were truly wise. As Aristotle states in his *Nichomachean Ethics*:

"The just, then, is an intermediate, since the judge is so. Now the judge restores equality; it is as though there were a line divided into unequal parts, and he took away that by which the greater segment exceeds the half, and added it to the smaller segment...But in associations for exchange this sort of justice does hold men together-reciprocity in accordance with a proportion and not on the basis of precisely equal return."

They thought the best way to avoid discord was for each citizen in a city-state to perfect their character or set of behavioral adaptations. This would optimize the fitness of each citizen to perform their role in society and relationships. Core to this perfected character were four artificial adaptive traits: 1) wisdom, 2) justice), 3) temperance, 4) courage.

If all citizens achieved the perfected character relative to their roles in life, then injustice and accompanying discord would be avoided. Then, if discord did occur, it would be resolved by relative justice. This would create and maintain a relative artificial equilibrium within a city-state.

This is a variation of the relative natural equilibrium in natural ecosystems. The Greeks must have observed this relative natural equilibrium in natural ecosystems. To the Greeks, those natural ecosystems were creations of the gods. Therefore, it was considered godlike for an individual or society to achieve such relative artificial equilibrium. Surely only peace and prosperity could be the result of homo sapiens achieving such godlike perfection.

10.7.5 ARTIFICIAL PERFECTION

"For the man [artist] who is truly good and wise, we think, bears all chances life becomingly and always makes the best use of [perfects] the army at his command and a good shoemaker [artist] makes the best shoes [perfects] out of the hides that are given him; and so with all other craftsmen [artists]."

Aristotle, "Nichomachean Ethics"

In this passage, Aristotle describes artificial perfection. He discusses both artificial behavioral adaptation (use of an army) and artificial structural adaptation (a shoe). Aristotle states that an artist's purpose is to perfect the art relative to the constraints within which he operates. A commander perfects the art of war (i.e., strategy, tactics, etc.) relative to the operational situation. The shoemaker perfects the production of his art relative to the raw materials he has available to him.

This is a core natural evolutionary concept. Each species perfects its phenotype through the process of adaptation relative to its natural ecosystem constraints. This includes both structural and behavioral adaptations. Products of art, the weapons and the tactics of a warrior, should be perfectly adapted to the constraints of each opponent and battlefield.

10.7.6   NATURAL & ARTIFICIAL INSTINCTS

"Neither by nature [natural evolution], then, nor contrary to nature [natural evolution] do the virtues [natural instincts] arise in us; rather we are adapted by nature [possess the natural adaptation of imagination] to receive them, and are made perfect by habit [artificial instincts]."

Aristotle, "Nichomachean Ethics"

Character was the combined set of instincts a human being possessed. The term "virtue" is really an instinct, artificial or natural. This is an adaptive trait in evolutionary terms. The perfected character or set of instinctual behaviors made a human fit for their specific role within a city-state.

Both Plato and Aristotle believed that four artificial behavioral adaptations were necessary for a perfected human character in the context of a city state: 1) wisdom, 2) justice, 3) temperance, 4) courage. These artificial instincts are still required for this purpose to this day.

In fact, the artificial instincts of justice and temperance were essential. These two artificial instincts enable a human to self-restrain natural instinctual behaviors. Homo sapiens must leverage their imagination to identify and modify destructive artificial instincts (habits).

If homo sapiens could perfect their artificial instincts through training, habituation, or imagination then they could become perfected and godlike. If humans and society achieved this godlike quality, then as Aristotle states in the *Nichomachean Ethics,* harmony would result:

"Another belief which harmonizes with our account is that the happy man lives well and does well; for we have practically defined happiness as a sort of good life and good action. The characteristics that are looked for

in happiness seem also, all of them, to belong to what we have defined happiness as being...With those who identify happiness with virtue or some one virtue our account is in harmony; for to virtue belongs virtuous activity."

10.7.7 ARTIFICIAL ADAPTATION

"...happiness [perfection] seems, however, even if it is not god-sent but comes as a result of virtue [imagination] and some process of learning or training, to be among the most godlike things; for that which is the prize and end of virtue [adaptation] seems to be the best thing in the world, and something godlike and blessed."

Aristotle, "Nichomachean Ethics"

In this passage, Aristotle describes artificial adaptation. The pattern of excellence revealed in the *Meno* is fundamentally about the process of artificial adaptation. By using our imagination we can develop new artificial structural and behavioral adaptations – thereby removing the constraints of our natural phenotype.

The use of imagination to artificially adapt is humanity's ultimate evolutionary advantage. To achieve optimum fitness by perfecting ourselves to our environment is considered godlike by the Greeks. As Aristotle says above, "the best thing in the world". By which he means evolutionary perfection.

10.7.8 ARTIFICIAL AND NATURAL ADAPTIVE TRAITS

"For in speaking about a man's character [set of adaptations] we do not say that he is wise [artificial adaptive trait] or has understanding but that he is good-tempered or temperate [artificial adaptive trait]; yet we praise the wise [artificial adaptive trait] man also with respect to this state of mind; and of states of mind we call those which merit praise virtues [adaptive traits]."

Aristotle, "Nichomachean Ethics"

In this passage, Aristotle is describing what comprises a human's character or set of behavioral adaptations. This includes both artificial and natural behavioral adaptations in the form of instincts. Those instincts that are beneficial to a human are classically named "virtues". Virtues are both natural and artificial instincts which are considered adaptive traits. Adaptive traits provide selective evolutionary value.

Imagination and wisdom are critical to developing and habituating artificial instincts. Imagination enables homo sapiens to see patterns of behavior of both you and of others. Wisdom is the capacity to discern what is beneficial and what is not. Both were necessary to perfect one's character.

10.8   RHETORIC – ARISTOTLE

10.8.1   INTRODUCTION

"Aristotle's Rhetoric is, if perhaps not the first, then at least the most important treatise on rhetoric, and is remembered also for its insightful psychology of the emotions. It exerted an influence on Roman orators such as Cicero and Quintilian."

Neel Burton, "Gang of Three"

Aristotle's *Rhetoric* is the ancient Greek treatise that is the basis for the art of homo sapiens rhetoric. Rhetoric is the art of persuading others to artificially select something: an idea, a political candidate, art, a course of action, etc. In *Rhetoric* Aristotle lays out a basic system of rhetorical argument that has served as the foundation for the art of rhetoric itself. It is considered the most important treatise on persuasion in history.

Aristotle established that three elements were core to the art of rhetoric: ethos, pathos, and logos. Ethos is the reputation regarding a person's character. Pathos is an appeal to a person's emotions based on their existing value, opinions, prejudices, etc. Logos is an appeal to a person through logical reasoning. Rhetoric is the art of combining these three elements in a way that successfully persuades an audience, a jury, an electorate, a consumer, etc. to artificially select your argument (art form).

10.8.2 CONCEPTUAL SELECTION

"On the other hand, the better the selection one makes of propositions suitable for special Lines of Argument, the nearer one comes, unconsciously, to setting up a science that is distinct from dialectic and rhetoric."

Aristotle, "Rhetoric"

In this passage, Aristotle describes conceptual selection. Conceptual selection is the process by which one selects a pattern (e.g., concept, set of concepts, etc.) for the design of art. This process is central to exercising our ultimate adaptive trait – imagination. A homo sapiens imagines the mental image of the art which is to be created. Then selects the concepts to be used to produce that art form. Below, Aristotle describes this selective process in *Rhetoric*:

"There is another method open to both calumniator and apologist. Since a given action can be done from many motives, the former must try to disparage it by selecting the worse motive of two, the latter to put to better construction on it."

Aristotle is providing instructions to the reader of *Rhetoric* on how to select the right concept in the form of motive. He is saying that a motive is a concept which must be applied effectively in the larger pattern (conceptual model) of your rhetorical argument. This is how motive is utilized within the conceptual model of the rhetorical argument. This will influence your audience's decision (artificial selection).

10.8.3 ARTIFICIAL SELECTION

"In a political debate the man who is forming a judgement is making a decision about his own vital interests...It is clear, then, that rhetorical study , in its strictest sense, is concerned with the modes of persuasion."

Aristotle, "Rhetoric"

In this passage, Aristotle is describing the process of artificial selection. This is the process by which a human being consciously and intentionally selects existing art forms (i.e., a restaurant meal, a candidate, a car, a political policy, etc.). That is what the art of rhetoric is about – influencing the artificial selection of others.

This was critically important in the Athenian assembly as each citizen had one vote – a white pebble for "yes" or a black pebble for "no". It was a direct form of democracy, so orators made an appeal to each individual citizen's self-interest.

The argument an orator makes is a form of art. Each orator would deliver his speech or variation in the Athenian assembly for his peer's judgment. Typically, multiple orators presented conflicting courses of action.

Each citizen present would then select or vote for the course of action (variation) they thought best for themselves and/or the city writ large. In a literal sense, they selected a specific orator's art or chose none. This is the process of artificial selection.

10.8.4 ARTIFICIAL ADAPTATION

"People always think well of speeches adapted to, and reflecting, their own character: and we can now see how to compose our speeches so as to adapt both them and ourselves to our audiences."

Aristotle, "Rhetoric"

In this passage, Aristotle is describing the process of artificial adaptation. In Rhetoric, he instructs the reader to adapt speeches (i.e., conceptual patterns) to be most fit in persuading their audiences. In effect, Aristotle is stating that the orator is searching for the argument most perfected to the specific audience's prejudices, opinions, value, etc. – maximizing the argument's probability of selection by the audience.

10.9 THE REPUBLIC – PLATO

10.9.1 INTRODUCTION

"Just as Plato leant upon Heraclitus' flux for his conception of the sensible world, so he leant upon Parmenides' unity for his conception of the intelligible world, which he rendered as the ideal, immutable realm of the

Forms...The Republic is often regarded as Plato's magnum opus. It has been voted the greatest work of philosophy ever written..."

Neel Burton, "The Gang of Three"

Plato's *Republic* is a Socratic dialogue similar to the *Meno*. The subject of the *Republic* is the practical organization of an ancient Greek city-state. It is Plato's most famous work and is the most read book in academia. The *Republic* is a foundational work of ancient Greek philosophy and has been tremendously influential in political theory.

In effect, the *Republic* is an exercise in applying evolutionary theory in the organization of human society. Almost all the patterns in evolution we discussed in chapter six are present. The ideas expressed in the book are a mix of natural and artificial evolutionary concepts. Plato systematically adapts these evolutionary patterns to the ideal pattern for an ancient Greek city state. Basically, Plato is an evolutionary scientist that focuses on one specific species – homo sapiens.

10.9.2 ARTIFICIAL SPECIES

"A State [artificial species], I said, arises, as I conceive, out of the needs of mankind [species]; no one is self-sufficing, but all of us have many wants. Can any other origin of a State [artificial variation] be imagined?... There can be no other...Then, I said, let us begin and create in idea a State [artificial species]; and yet our true creator is necessity, who is the mother of our invention."

Plato, "The Republic"

In this passage, Plato introduces the concept that artificial societies function as types of artificial species. We have artificially organized into different types of societies (e.g., tribes, nations, empires, city-states, etc.) for millennia. Over time, these artificial societies slowly evolved into distinct cultures. Culture is the umbrella term for a human society's social behavior, institutions, norms, beliefs, arts, laws, etc.

This is how the ancient Greeks viewed non-Greeks. They viewed people who spoke Greek as belonging to their culture and those who people who did not as barbarians. This cultural distinction classified non-Greeks as artificially different than Greeks. Since in fact we are all genetically the same species, the distinction is purely an artificial one.

The cultures (i.e., artificial species) that homo sapiens establish engage in survival of the fittest contests similar to natural evolution. Examples are Macedonia vs. Persia, Rome vs. Carthage, and Byzantine vs Ottoman. These artificial societies each engaged in a different survival of the fittest contest that resulted in the artificial extinction of one of them. The losing cultures became rare and eventually became extinct. In the *Republic* Plato is attempting to leverage evolutionary concepts to construct a city state that would survive such competition.

10.9.3 NATURAL EXPANSION

"And the country which was enough to support the original inhabitants will be too small now, and not enough?...Quite true...Then a slice of our neighbors' land will be wanted by us for pasture and tillage, and they will want a slice of ours, if, like ourselves, they exceed the limit of necessity, and give themselves up to the unlimited accumulation of wealth?...That, Socrates, will be inevitable...And so we shall go to war, Glaucon. Shall we not?...Most certainly, he replied."

Plato, "The Republic

In this passage, Plato describes natural evolutionary expansion. It is literally the same concept as described by Charles Darwin in *On the Origin of Species*. The species homo sapiens wants to expand the same as any other species. As we reach the limit of our resources (e.g., territory, water, food, etc.) the survival of the fittest context inevitably occurs. That is typically the underlying cause in most homo sapiens conflicts.

It is this concept of natural expansion for which Plato seeks a solution. Plato hoped to control our natural instincts to expand socially, economically, sexually, etc. He developed a set of artificial adaptations (i.e., virtues, belief system, etc.) – that restrain our natural instincts.

It is the instinct to expand that inevitably destabilizes the relative equilibrium established within a society. Plato believed that the restraint of this instinct can only be accomplished by the rulers of the state deceiving its own citizens.

This is why Plato believes only the rulers should be allowed the privilege of lying for the good of the state.

10.9.4 ARTIFICIAL SELECTION

"Then there must be a selection [artificial selection]. Let us note among the guardians those who in their whole life show the greatest eagerness to do what is for the good of their country, and the greatest repugnance to do what is against their interests...Those are the right men."

Plato, "The Republic"

In this passage, Plato describes the process of artificial selection. The art being selected is the artificial phenotype of a particular guardian. If the guardian has the right artificial adaptations (e.g., justice, virtue, temperance, courage, etc.), then the rulers will artificially select him for higher office.

Technically it is a combination of both natural adaptations and artificial adaptations being selected since homo sapiens possess a natural phenotype. But it is an intentional selection by a homo sapiens, so it is not natural selection, but artificial selection.

10.9.5 SELECTIVE BREEDING

"Why, I said, the principle has been already laid down that the best of either sex should be united with the best as often, and the inferior with the inferior, as seldom as possible...if the flock is to be maintained in first-rate condition. Now these goings on must be a secret which the rulers only know, or there will be further danger of our herd, as the guardians may be termed, breaking out into rebellion."

Plato, "The Republic"

In this passage, Plato applies Charles Darwin's evolutionary concept of selective breeding. It is literally identical. However, the species he is breeding in the *Republic* is homo sapiens. As dog breeders attempt to do today, Plato hoped to dictate the reproduction patterns of a population. By doing so, he hoped to produce offspring with the most beneficial natural adaptations (i.e., strength, beauty, imagination, intelligence, etc.).

However, selective breeding is an artificial form of sexual selection. If the breed produced by this artificial sexual selection is beneficial to homo sapiens, it will be artificially selected for human activity (e.g. a fast horse, excellent milk producing cow, etc.). If this is not the case, then that variation will not be artificially selected. It will most likely not be selected for the process of selective breeding.

10.9.6 ARTIFICIAL EQUILIBRIUM

"But the latter be the truth, then the guardians and auxillaries, and all others equally with them, must be compelled or induced to do their own work in the best way. And thus the whole State [artificial species] will grow up in a noble order [artificial equilibrium], and the several classes will receive the proportion of happiness which nature assigns to them."

Plato, "The Republic"

In this passage, Plato discusses the concept of relative artificial equilibrium described earlier in the *Nichomachean Ethics*. He applied some of those evolutionary concepts to establish and sustain the ideal ancient Greek city-state.

10.9.7 ARTIFICIAL PERFECTION

"I was only going to ask whether, if we have discovered them, we are to require that the just man should do nothing fail of absolute justice; or may we be satisfied with an approximation, and the attainment in him of a higher degree of justice than is to be found in other men?... The approximation will be enough...We are enquiring into the nature of absolute justice and into the character of the perfectly just, and into injustice and the perfectly unjust, that we might have an ideal."

Plato, "The Republic"

In this passage, Plato discusses the concept of artificial perfection. This is the same concept Aristotle described in the *Nichomachean Ethics*. Plato discusses how the adaptive trait of justice must be perfected in homo sapiens' affairs. He believes justice must be applied not absolutely but in a relative way.

To establish and sustain the artificial equilibrium of a Greek city-state, justice must be applied relative to the specific situation. The artificial adaptation of justice must be exercised in a manner that ultimately achieves prosperity. This is how an adaptation in nature is perfected relative to its' natural ecosystem.

10.10 CONCLUSION

"So erst the Sage [Pythagoras] with scientific truth...In Grecian temples taught the attentive youth; With ceaseless change how restless atoms pass [Democritus]...From life to life, a transmigrating mass...How the same organs, which to-day compose...The Poisonous henbane, or the fragrant rose...May with to-morrow's sun new forms compile...Frown in the Hero [Socrates], in the Beauty smile [Plato]...When drew the breath enlighten'd Sage [Aristotle] the moral plan...That man should ever be the friend of man...Should eye with tenderness all living forms...His brother-emmets, and his sister-worms."

Erasmus Darwin, "To the Stars"

The ancient Greek philosophers were evolutionary scientists. They utilized homo sapiens' most powerful natural adaptation – imagination – to discover the pattern of evolution. However, the ancient Greek would have called it "the pattern of the gods in nature".

Charles and Erasmus partially adopted the ancient Greek point of view. This enabled them to perceive the conceptual patterns of evolution in ancient Greek philosophy. They then mined ancient Greek texts to assemble the underlying conceptual pattern. It was this underlying pattern that Erasmus and Charles leveraged to develop the theory of natural evolution.

However, selective breeding is not artificial selection. Selective breeding is just a form of sexual selection – albeit an artificial one. This means that artificial selection is not part of the pattern of natural evolution. It appears that artificial selection is part of an entirely different pattern the ancient Greek philosophers discovered. Given these facts, we need a new definition for the concept of artificial selection.

11

CHAPTER 11: ARTIFICIAL SELECTION REDEFINED

"We meet with this admission in the writings of almost every experienced naturalist; or as Milne Edwards has well expressed it, Nature is prodigal in variety, but niggard in innovation. Why, on the theory of Creation, should there be so much variety and so little real novelty?"

Charles Darwin, "On the Origin of Species"

11.1   INTRODUCTION

"But if on the other hand art imitates nature, and it is the part of the same discipline to know the form and the matter up to a point (e.g. the doctor has knowledge of health...and the builder both of the form of the house and of the matter, namely that it is bricks and beams, and so forth): if this is so, it would be the part of physics also to know nature in both its senses."

Aristotle, "Physics"

Artificial selection is an anomalous concept in Charles Darwin's theory of natural evolution. We now know why that is the case. Artificial selection was a concept Charles Darwin mined from the pattern from ancient Greek Philosophical texts.

Selective breeding was the use case that Charles Darwin selected from the Greek pattern. He used the process of selective breeding to frame his definition for the term artificial selection. Charles did so as a rhetorical device at the beginning of his book *On the Origin of Species.*

Charles hoped that by starting his pattern of natural evolution with concepts homo sapiens already intimately understood, it would lend credibility to his argument. His rhetorical device worked exactly as he had intended. So, we will pick selective breeding as our logical starting point for working out the true definition of artificial selection.

11.2   SELECTIVE BREEDING

"Why, I said, the principle has been already laid down that the best of either sex should be united with the best as often, and the inferior with the inferior, as seldom as possible...if the flock is to be maintained in first-rate condition. Now these goings on must be a secret which the rulers only know, or there will be further danger of our herd, as the guardians may be termed, breaking out into rebellion."

Plato, "The Republic"

We have already covered selective breeding in chapter seven. But it is now clear Charles Darwin's understanding of the process was not accurate. Here we will reconsider what selective breeding fundamentally is and how it works.

Selective breeding is the process by which homo sapiens interferes in the natural reproduction of other species (e.g., dogs, cattle, wheat, rice, etc.). We will select the use case of dog breeding to conduct our analysis. Let us now consider exactly how the process of dog breeding works.

The breeder has an end in mind for the mature dogs produced via natural reproduction. Let us select the of hunting foxes as our ***end*** for this exercise. The foxhound is the variation of the species dog bred for such a purpose. The foxhound is the ***means*** to hunt foxes in this use case.

What makes a good hunter of foxes – the ability to track the scent of foxes through any terrain. This combines a keen sense of smell (structural adaptation – nose organ) with a tenacious energy and drive (behavioral adaptation – instinct). Plus, it must have a gentle temperament (behavioral adaptation – instinct) easy for homo sapiens to control during the hunt.

So, that is the natural phenotypic pattern homo sapiens bred foxhounds for – 1) keen sense of smell, 2) tenacious energy and drive, 3) gentle temperament. But a homo sapiens first had to conceive that pattern or intended **_means_** in his or her mind.

Each of the three phenotypic traits has a separate pattern or concept in the context of hunting foxes. This is the same as the different concepts which comprise a product (i.e., cell phone, car, etc.) which you purchase at the store. We identified an ancient Greek concept for this activity in the last chapter – conceptual selection.

11.3   CONCEPTUAL SELECTION

"The arts, therefore, which govern the matter and have knowledge are two, namely the art which uses the product and <u>the art which directs the production of it</u>...For the helmsman knows and prescribes what sort of form a helm should have, the other from what wood it should be made and by what means of what operations. In the products of art, however, we make the material with a view to the function, whereas in the products of nature the matter is there all along."

Aristotle, "Physics"

Conceptual selection is "the art which directs the production of" the art. A breeder must first select the pattern, such as the foxhounds, which he wants to produce. But as the British writer Thomas Carlyle once stated:

"Thought is the parent of the deed."

The breeder first **searches** for the components of the pattern he wants to produce. He then **selects** this pattern in his or her mind which **conceives** a new pattern. Then the breeder mates the two natural organisms in hopes of naturally **conceiving** the previously intellectually **conceived** pattern.

11.4   PRODUCTION OF ART

"Things which are formed in nature are in the same case as these products of art...The natural things which (like the artificial objects previously considered) can be produced spontaneously are those whose matter can

be moved by itself in the way in which the seed usually moves it; those things which have not such matter cannot be produced except from the parent animals themselves."

Aristotle, "Metaphysics"

Selective breeding leads to the development and production of natural offspring. In our example the pattern desired is the natural phenotype of a foxhound. Specifically, it is a foxhound with the perfect expression of three adaptations the breeder desires. It becomes a foxhound perfected as a ***means*** to the ***end*** of hunting foxes.

But the foxhound offspring produced by selective breeding is not a result of sexual or natural selection. Homo sapiens interferes to produce a result beneficial to us – not necessarily to the dog. The result is an artificial one since homo sapiens interferes. This makes the foxhound offspring produced by selective breeding art – making selective breeding an art form.

The fact that the process of natural reproduction is utilized doesn't matter. The pattern being produced is an artificial one conceived by a homo sapiens' mind – not by the preferences of a female foxhound. Also, since we keep them as pets, homo sapiens determine the natural fitness of the species dog. The species now have artificial fitness determined by the usefulness to homo sapiens. This makes it equivalent to any other art produced by homo sapiens. The more useful we find a type of art, the more artificial fitness that art possesses in human activity.

11.5   ARTIFICIAL SELECTION

"The arts, therefore, which govern the matter and have knowledge are two, <u>namely the art which uses the product</u> and the art which directs the production of it...For the helmsman knows and prescribes what sort of form a helm should have, the other from what wood it should be made and by what means of what operations. In the products of art, however, we make the material with a view to the function, whereas in the products of nature the matter is there all along."

Aristotle, "Physics"

The produced art or foxhound offspring may or may not be the **_means_** to the **_end_** of fox hunting. The process of natural reproduction isn't an exact science. Natural evolution intentionally varies as part of the reproductive process. So, it may take multiple natural reproductive cycles before the perfected foxhound phenotype is produced.

Not every foxhound puppy (i.e., art) will possess the three adaptations the breeder desires. Some puppies will not possess one of the adaptations to a sufficient degree. Homo sapiens must select the art (foxhounds) produced for use in human activity. Art is an artificial production. So, the breeder is selecting an artificial "thing". A puppy born lame will not be selected for fox hunting. This is the true definition of artificial selection:

Artificial Selection is the selection of art produced
by homo sapiens for use in human activity

11.6 CONCLUSION

"But if on the other hand art imitates nature, and it is the part of the same discipline to know the form and the matter up to a point (e.g. the doctor has knowledge of health...and the builder both of the form of the house and of the matter, namely that it is bricks and beams, and so forth): if this is so, it would be the part of physics also to know nature in both its senses."

Aristotle, "Physics"

Artificial selection is not a part of the pattern of natural evolution. It is part of a pattern that is a variation of natural evolution. Charles Darwin has given us the clue as to what artificial selection is in the excerpt below from his book _On the Origin of Species_:

"We meet with this admission in the writings of almost every experienced naturalist; or as Milne Edwards has well expressed it, Nature is prodigal in variety, but niggard in innovation. Why, on the theory of Creation, should there be so much variety and so little real novelty?"

Artificial selection is a variation of natural selection. Natural selection is the core mechanism of natural evolution. If artificial selection is a variation of natural selection, then there must be an entirely different pattern of evolution – an artificial pattern. We uncovered some of these artificial conceptual variations when we decoded ancient Greek philosophical texts. It is the pattern in human activity the family Darwin missed.

12

CHAPTER 12: ARTIFICIAL EVOLUTIONARY PATTERNS

"For the word 'nature' is applied to what is according to nature and the natural in the same way as 'art' is applied to what is artistic or a work of art."

Aristotle, "Physics"

12.1 ARTIFICIAL ECOSYSTEMS

"This book presents a comprehensive framework of analytical techniques to help a firm analyze its industry [artificial ecosystem] as a whole and predict the industry's future [artificial] evolution, to understand its competitors and its own position, and to translate this analysis into a competitive strategy for a particular business."

Michael E. Porter, "Competitive Strategy"

An artificial ecosystem is a complex web of interdependencies created by patterns of artificial coevolution, artificial co-adaptation, and natural environments. Each thread of the web possesses many-to-many relationships. This creates a set of tangled relationships between artificial organisms and the environment. In addition, this web is not static, but a moving target due to dynamic change.

In his book *Competitive Strategy*, Michael Porter provides analysis of persistent patterns in commercial industries. A commercial industry is an artificial variation of ecosystems found in natural evolution. That is why Thomas Siebel was able to accurately pattern-match a variation of natural punctuated equilibrium to an extreme form of digital disruption.

Siebel also applied this concept variation to Nation State competition as well. This is because collectively the world's sovereign states comprise an international artificial ecosystem. Each of these sovereign states is an artificial ecosystem dominated by one or more artificial species. In addition, there are sub ecosystems within this highest-level artificial ecosystem such as Europe, the Pacific, Africa, etc.

Artificial ecosystems exist in all aspects of human activity. Examples include international regions, nation states, political parties, university systems, commercial industries, professional sports leagues, etc. We artificially develop the pattern and then artificially implement the pattern. All the listed human ecosystems are products of human art and are therefore artificial.

12.2   ARTIFICIAL EVOLUTIONARY CONDITIONS

"The number and magnitude of challenging events are increasing, and the environment in which today's organizations operate is often described as volatile, uncertain, complex, and ambiguous (VUCA)."

Axelos Global Best Practice, "ITIL 4: Digital and IT Strategy"

Artificial evolutionary conditions are a variation of natural evolutionary conditions. However, artificial evolutionary conditions can be the result of a combination of natural and artificial evolutionary patterns. An example of this is the combined impact on an artificial ecosystem of both climate and technological change. Changes in climate are natural evolutionary disruptions to human activity. Rapid technological change, digital disruption, causes artificial volatility in human activity. When both happen simultaneously, it often causes unpredictable events to occur.

A potential scenario for this phenomenon is conceivable in the U.S. insurance industry. Currently, Florida is being impacted by changes in climate. As a result insurers are withdrawing from the Florida insurance market. They see Florida's extreme and volatile weather patterns as unacceptable risks for their nation-wide business model.

Then imagine a new technological innovation disrupts the entire insurance industry. This disruption would also affect the Florida insurance market. As a result, the Florida insurance market would be disrupted by a combination of natural and artificial evolutionary conditions. This could occur in different variations in multiple states simultaneously. Ultimately, this dynamic will shape the overall artificial ecosystem which is the nation-wide insurance industry.

All the natural evolutionary conditions occur as variations in artificial ecosystems. Axelos and Oxford College publications described earlier have already confirmed this to be the case. They described the identical conditions in which natural species must compete as described by Charles Darwin's in his book *On the Origin of Species.*

12.3 ARTIFICIAL EVOLUTIONARY SPACES

12.3.1 ARTIFICIAL COMPETITIVE SPACE

"Section 2. The House of Representatives shall be composed of members chosen every second year by the people of the several states, and the electors in each state shall have the qualifications requisite for electors of the most numerous branch of the state legislature...Section 4. The times, places and manner of holding elections for Senators and Representatives, shall be prescribed in each state by the legislature thereof; but the Congress may at any time by law make or alter such regulations, exact as to the places of choosing Senators."

We The People, *"The Constitution of the United States of America"*

The Founding Fathers of the United States of America established a new constitutional order. Part of this new constitutional order was the creation of federal elections. The U.S. Constitution requires recurring elections be conducted for the U.S. Congress and the Office of the President. Candidates must compete for the votes of the people to attain federal office.

An election is really an artificial competitive space. Votes are scarce resources to the political candidates. Candidates compete to be artificially selected by voters at the end of each election. It is an artificial survival of the fittest contest with the winner taking all. As in nature, the competition is typically fierce. U.S. Presidential Election battleground states are competitive spaces every four years.

There are many other artificial competitive spaces. Examples include commercial industries, beauty pageants, sports leagues, academic theories, etc. All these spaces of homo sapiens competition are creations of art not found in nature. We create these competitive arts to channel our instinctive need to compete and expand.

An artificial competitive space is where competition actively occurs

12.3.2 ARTIFICIAL CONFINED SPACE

"Under the agreement, Rozelle would be the commissioner of both leagues; the leagues would play an annual world championship game beginning after the 1966 season...After much political wrangling...On October 21, the now-forgotten tax bill and pro football's antitrust exemption became law...The issue was settled. Realignment was complete. The merger was done...The war was over."

Joe Horrigan, "NFL Century"

The American football industry [an artificial ecosystem] experienced a period of intense competition during the mid-twentieth century. At that time the American football industry was open to the entry of new competitors. This created an evolving competitive dynamic within the industry.

The American Football League (AFL) and the National Football League (NFL) were in open and fierce competition over the consumers of the industry [artificial territory]. This competition resulted in the merger of the AFL and NFL into a single league – today's NFL.

This restricted the entry of any new franchises into the combined league without formal authorization by the NFL and its owners. This established an artificially confined space for competition between the new league's teams.

This is how artificial ecosystem boundaries are created. Homo sapiens conduct the process of conceptual selection and produce new art. In the case of the NFL, the new art was the merger agreement. Other forms of art that create artificial ecosystems include contracts, charters, armistice, laws, etc. The created art restricts human competition artificially until one or more parties violates the artificial boundary the new art produced.

> ### *An artificial confined space is surrounded by*
> ### *artificial barriers that restrict competition*

12.3.3 ARTIFICIAL DOMINANCE SPACE

"Largely as a result of his keen insights, his many examples, and his rhetorical flourish in crafting On the Origin of Species, Darwin is now regarded as the giant modern evolutionary theory, his name virtually synonymous with evolution by means of natural selection."

Vassiliki Betty Smocovitis, "The Theory of Evolution"

Artificial ecosystems are inhabited by a diverse set of conceptual species. These conceptual species compete intellectually just as species do in natural evolution. As homo sapiens conceptually select these conceptual species, some will expand while others will contract in terms of use.

It is a zero-sum game, given the limited number of homo sapiens on the planet. Those that expand the most relative to their competitors will have achieved dominance. This dominance exists in the minds of homo sapiens' populations. The dominant concepts exert control and influence over other conceptual species' capacity to expand. This makes the minds where a concept has achieved dominance a conceptual dominance space.

Charles Darwin's idea of the theory of evolution by the means of natural selection has stood the test of time. He has won the intellectual survival of the fittest contest within the academic community. Charles Darwin's idea expressed in *On the Origin of Species* has achieved intellectual dominance. The field of evolutionary biology is Charles Darwin's conceptual species' dominance space. It will continue to dominate until a new conceptual species arises to displace it.

Artificial dominance spaces include many variations. The evolutionary biology academic field, or artificial ecosystem, is one example. Others include Standard Oil's dominance of the U.S. oil industry, the Roman Empire's dominance of the Mediterranean, and the New England Patriot's past dominance of the NFL. These are all variations like Charles Darwin's artificial dominance of evolutionary biology.

12.3.4 ARTIFICIAL NEUTRAL SPACE

"But that situation would not have been very different from what finally emerged along the 38th parallel…In spring of 1951, a new American offensive under General Ridgeway was grinding its way north…It had liberated Seoul and crossed the 38th Parallel when, in June 1951, the communists proposed armistice negotiations…a painful equilibrium emerged between China's physical limitations and America's psychological inhibitions…The stalemate America sought descended on both the military and diplomatic fronts."

Henry Kissinger, "Diplomacy"

Artificial ecosystems are spaces of fierce competition over scarce resources. Often two artificial species will establish two mutually exclusive dominance spaces in artificial ecosystems. There then develops a buffer zone, between these two artificial species. In this buffer zone neither artificial species exerts control or influence over the other.

This space, whether physical or artificial, where neither of the two dominant artificial species exists, is an artificial neutral space. This artificial neutral space is created by some form of art. An example is the Korean peninsula's demilitarized zone across the 38th Parallel. This artificial neutral space is comprised of military mines, barriers, troops, etc. on either side of a strip of land. However, the strip of land in between is considered a "no man's land" where neither artificial species asserts dominance. Other examples of artificial neutral spaces are commercial market niches, U.S. political election independent voters, etc.

An artificial neutral space is a space where there
is no clear dominant artificial species

12.3.5 ARTIFICIAL OPEN SPACE

"Cyberspace is a global fabric, with few tools to define jurisdictional boundaries...we are situated in a social system, we are inevitably and invariably anchored in a natural environment [natural ecosystems] for the provision of life-supporting properties, and, increasingly, we find ourselves engaged in cyberspace, the built environment [artificial ecosystems]... It may also be difficult to conceive of a world enabled by other than an integrated joint cyber-IR system."

Nazli Choucri & David D. Clark, "International Relations in the Cyber Age"

Artificial ecosystems have varying environments across human activity. Some are relatively small due to artificial and/or natural barriers (i.e., confined), and others range widely. Those artificial ecosystems which are open to the introduction of new human activity (i.e., no artificial barriers) are considered open space.

Within open spaces, competition is more intense and dynamic as humans are exposed to increasing competition over time. The homo sapiens who rise to dominance in open spaces tend to develop the most competitive adaptations.

Cyberspace is a clear open space in human activity. A prime example of this is the hacking of the U.S. Democratic National Committee's servers housed in Washington, D.C. The hack was performed by staff of Russia's GRU seated in Moscow. The artificial highway of cyberspace enabled the GRU agents to artificially traverse the U.S.'s geographic and jurisdictional boundaries. The Russian cyber-attack on the 2016 U.S. election process would not have been possible in 1950. It is the 21st Century creation of cyberspace, which has created artificially open space across all human activity.

Artificial open space has no significant barriers to human competition

12.4   ARTIFICIAL ORGANISMS

"It would seem that in animals, just as in ships and things not naturally organized, that which causes motion is separate from that which suffers motion, and that it is only in this sense that the animal as a whole causes its own motion."

Aristotle, "Physics"

The word organism is derived from two Ancient Greek words. The first, 'organon', means instrument, tool, or organ. The second, 'ismus' (a word evolved thru Ancient Greek, Latin, and finally English) roughly means "a system or condition". This translates to an organism being a complete system of organs – that defines the core of an organism. Combined with other elements of an organism, this creates its structure. The ancient Greeks considered this to apply to both natural and artificial organisms.

There are many types of artificial organisms on earth. Every artificial organism is made of one or more artificial organs. Examples of artificial organisms are corporations, political committees, cars, planes, predator drones, trains, chariots, etc. Most artificial organisms must consume some resource (i.e., human labor, money, gas, electricity, sunlight, etc.) to sustain their system of artificial organs.

Homo sapiens must reproduce, adapt, and maintain artificial organisms' system of organs. To date most artificial organisms generally do not possess the capacity to independently respond to environmental stimuli. However, this situation is rapidly changing due to the advent of artificial intelligence.

12.5   ARTIFICIAL SPECIES

The concept of artificial species in evolutionary theory roughly defines a large group of similar objects. These objects typically possess the same observable pattern. The species concept provides artificial evolutionary theory, a rough working model for grouping objects. This object grouping aids homo sapiens in organizing and simplifying the complexity of our artificial reality. Examples of artificial species are art, human culture, and concepts.

12.5.1 ARTIFICIAL SPECIES – ART

"And thus the certainty of being able to exchange all that surplus part of the produce of his own labour, which is over and above his own consumption, for such parts of the produce of other men's labour as he may occasion for, encourages every man to apply himself to a particular occupation, and to cultivate and bring to perfection whatever talent or genius he may possess for that particular species of business."

Adam Smith, "The Wealth of Nations"

Homo sapiens generally establishes corporations (artificial organisms) to produce art in the form of products and services (i.e., artificial structural and behavioral adaptations). The products and services of each corporation are effectively variations of artificial phenotypic traits. Combined, the variations created by all the corporations appear to the consumer to be just one artificial species.

Homo sapiens consumers artificially select from these variations for their own evolutionary benefit (i.e., survival, reproduction, etc.). Adam Smith explicitly describes this concept as each industry is a "species of business". If a corporation's variation of a product or service is not artificially selected, then that product or service goes artificially extinct.

Some corporations or artificial organisms can evolve into new artificial species of business. Studebaker was one such corporation. The horse and buggy species of business was going extinct. So, Studebaker adapted to become a variation of a new species of business – automobiles.

Studebaker was able to escape the extinction of the old industry by evolving its product lines to compete in an emerging one. This type of dramatic and rapid adaptation is only possible in artificial evolution. Typically, natural species are too constrained by their genotype and phenotype to adapt quickly and dramatically to that degree.

12.5.2 ARTIFICIAL SPECIES – HUMAN CULTURE

"THE STRUGGLE BETWEEN Rome and Carthage spanned over a century from the first clash in 265 down to the final destruction of Carthage in 146...By the end of the conflict Carthage was in ruins, its life as a state

ended and its culture almost totally extinguished. Between 265 and 146 Rome rose from being a purely Italian power into a position of unrivalled dominance throughout the Mediterranean base..."

Adrian Goldsworthy, "The Fall of Carthage"

Homo sapiens are born with natural instincts to organize in hierarchies for cooperation. However, the competitive dynamic and physical environment for human societies vary across the planet. Therefore, homo sapiens develop artificial societal adaptations perfected to their specific ecosystems. We call the aggregated societal art created for human cooperation culture. Culture includes art such as knowledge, belief, customs, laws, norms, traditions, cuisine, etc.

The art of human cultures evolves over time into distinct cultures such as Chinese, British, Zulu, Egyptian, Iroquois, Indian, Aztec, etc. Homo sapiens develop governmental systems as art to organize the hierarchy across each culture. As a result, homo sapiens' organization resembles that of an ant colony with division of labor across the Nation State. Nation States often have a single dominant culture (e.g., English) and one or more subcultures (e.g., Welsh, Scottish, etc.).

Each Nation State then functions as an artificial species derived from the natural species homo sapiens. An example is the Punic Wars between Rome and Carthage. The Punic Wars began with limited competition over the territory and resources of the island of Sicily.

The Punic Wars then evolved into a struggle for existence between the two cultures with only one survivor – Rome. The Carthaginian culture, in all practical terms, went artificially extinct. Homo sapiens cultures follow the same lifecycle as species found in nature – speciate, vary, expand, stasis, contract, extinction or speciate again. The difference is that cultures do so artificially as they are a product of homo sapiens art.

This concept has already been identified by psychologist James Mark Baldwin. In 1896 he developed a theory known as "The Baldwin Effect". As Joseph Jebelli states in his book How the Mind Changed:

"As humans became increasingly reliant on culture [artificial species] changes [artificial adaptations] that enabled them to cooperate effectively – changing social patterns [artificial instincts], changing social technologies

CHAPTER 12: ARTIFICIAL EVOLUTIONARY PATTERNS

[structural adaptations], changing social goals – they became increasingly able to replace genetic evolution [natural evolution] with cultural evolution [artificial evolution]...They set us on a [evolutionary] trajectory of rapid technological innovation [accelerated artificial evolution] – the fruits of which , if tech experts are to be believed, will either save or destroy [cause extinction of] humankind."

Baldwin discovered the rough partial pattern of artificial evolution. He simply did not go far enough in terms of evolutionary thought. However, his concept is and will always be logically valid. He matched the pattern to the process of artificial evolution without conscious knowledge of that fact. This is a recurring pattern in our species' incrementally perceiving the pattern of artificial evolution in nature. We identify an individual pattern but lack the proper context to understand that it is part of a much larger pattern.

12.5.3 ARTIFICIAL SPECIES – CONCEPTS

"If genera are different and co-ordinate, their differentiae are themselves different in kind. Take as an instance the genus 'animal' and the genus 'knowledge'. 'With feet', 'two-footed', 'winged', 'aquatic', are differentiae of 'animal'; the species of knowledge are not distinguished by the same differentiae. One species of knowledge does not differ from another in being 'two-footed'."

Aristotle, "Categories"

Homo sapiens have assembled knowledge since our species first acquired the natural adaptation of imagination. This knowledge is preserved in the form of concepts or patterns humanity has scientifically discerned in nature. We classify these concepts into artificial species of related concepts just as in natural species. The academic departments of modern universities are like species kingdoms of homo sapiens concepts. At the lowest level of conceptual thinking are conceptual species and their conceptual variations.

Homo sapiens' conceptual species initially emerge in the form of ideas. Once the idea expands to most homo sapiens' minds, it becomes dominant. This intellectual dominance leads to an idea becoming a concept. Concepts continue to exist until a new idea emerges which threatens that existing concept.

New homo sapiens ideas often cause long held concepts to become obsolete or extinct in homo sapiens' thinking. This is the standard pattern which natural species follow in evolution. Concepts do so as well, but in an artificial sense. This is a variation of the speciation cycle found in natural evolution.

12.6 ARTIFICIAL EXPANSION

"Almost as if according to some natural law, in every century there seems to emerge a country with the power, the will, and the intellectual and moral impetus to shape the entire system in accordance with its own values."

Henry Kissinger, "Diplomacy"

Homo sapiens are born needing to expand socially and sexually. Both are inextricably linked within our behavior. This is due to the natural instincts encoded in our genes. The need to expand socially or achieve social status is survival based. Social expansion creates the conditions for sexual expansion. This means access to the best mates within our specific social group. It is mating that enables us to expand literally in the form of offspring. These two facts are the key to understanding humanity at every level – individual, relationship, group, societal, etc.

In homo sapiens this need to expand is expressed artificially through competition in various forms such as military, economic, social, etc. Cultures are driven to expand by the aggregate need of its individuals to expand in terms of territory, wealth, reputation, etc. Businesses are driven by the aggregate need of its individuals to expand in terms of monetary reward, reputation, and social status. Individuals are driven to expand in terms of social influence through social media platforms.

Charles Darwin published his book under time pressure to ensure his reputation expanded. He wanted to ensure he received the credit for discovering the theory of natural evolution. These are all artificial expressions of our natural instinct to expand.

12.7   ARTIFICIAL OFFSPRING

"But souls [minds] which are pregnant – for there certainly are men who are more creative in their souls [minds] than in their bodies – conceive that which is proper for the soul [mind] to conceive or contain...Who, when he thinks of Homer and Hesiod and other great poets, would not rather have their children [new ideas] than ordinary human ones? Who would not emulate them in the creation of children [new ideas] such as theirs, which have preserved their memory and given them everlasting glory?"

Plato, "Symposium"

Artificial offspring are the art produced by homo sapiens through the process of artificial inheritance. This art can take the form of artificial organisms (i.e., automobiles, planes, etc.), artificial adaptations (i.e., winter coats, lean six sigma, etc.), and conceptual frameworks (i.e., ITIL 4, republican form of government, etc.).

Plato has Socrates describing his ideas as his "children" as above. Plato meant this literally, not figuratively. The very words used to describe generating an idea, "conceive", and the result of the process, a "concept", are literal examples of how artificial offspring are produced. Everything produced in art begins as an idea or new pattern conceived in your imagination. When that new idea is transitioned to a form of art, this is artificially giving birth to a physical "thing". As we already covered, this is exactly how natural reproduction works – information (DNA) is transformed into a physical "thing" – a newborn homo sapiens child.

This new art is typically created by combining and adapting existing homo sapiens conceptual knowledge. This is like how natural inheritance leverages genes to produce new phenotypic traits in natural offspring. The difference is that artificial offspring can potentially be produced from concepts available to homo sapiens' across all existing knowledge. Natural reproduction is limited to producing natural offspring only with the genotype initially provided by both parents. This difference means artificial reproduction has the potential for unlimited adaptive capacity.

12.8 THE ARTIFICIAL STRUGGLE FOR EXISTENCE

"Rome did not tolerate a former enemy existing as anything more than a clearly subordinate ally...The disaster at the Caudine Forks in 321 BC was the last time that Rome accepted peace as the clear loser in a war. The fourth-century BC struggle against the expanding hill peoples of Central Italy was certainly bitter and may have encouraged the Romans to think of war as a struggle for their very existence."

Adrian Goldsworthy, "Roman Warfare"

Artificial species, artificial organisms, concepts, and art are in an artificial struggle for existence. Homo sapiens' cultures, corporations, long-held concepts, and artificial adaptations/organisms are constantly being selected through competition. This competition is happening in the context of the natural struggle for existence. Homo sapiens' warfare is an area where the natural and artificial struggles for existence both occur simultaneously. However, the two concepts are logically distinct.

Human cultures are naturally selected through natural evolutionary competition (i.e., warfare). The organisms that comprise a culture are natural beings. Each culture functions as an artificial species within the natural species homo sapiens.

As with Rome and Carthage, the cultures often enter a survival of the fittest contest with only one eventual winner. In this case, Carthage's culture virtually ceased to exist. However, the cultures that win gain natural evolutionary benefits. Victory by a specific culture gives its homo sapiens enhanced ability to survive and reproduce.

In the case of corporations, concepts, and artificial adaptations/organisms the competition is purely artificial. The artificial selection performed by human beings determines which artificial adaptations and/or organisms survive or go extinct. Cell phones caused payphones to all but become extinct. Automobiles caused the horse and buggy to become extinct. The Republican Party caused the extinction of the Whig Party in American politics. Charles Darwin's theory of natural evolution by natural selection eventually caused the extinction of competing theories.

12.9   ARTIFICIAL ADAPTION

"Truly adaptive firms with adaptive cultures are awesome competitive machines. They produce superb products and services faster. They run circles around bloated bureaucracies. Even when they have far fewer resources and patents or less market share, they compete and win again and again."

John Kotter, "Leading Change"

In evolutionary competition, homo sapiens constantly had to change and expand to survive. In artificial evolution, there are two main ways homo sapiens can change in direct response to competitive or environmental conditions. The first, artificial adaptations, are new or modified art produced via the process of artificial inheritance. The second, artificial phenotypic plasticity, does not require the use of the process of artificial inheritance. They are distinct.

12.9.1   THE PROCESS OF ARTIFICIAL ADAPTATION

For simplicity's sake, we will use the term ***artificial adaption*** to describe the process of artificial adaptation in this book. Artificial adaption is the mechanism by which homo sapiens modify art in response to competitive and/or environmental pressures. This art is produced through the process of artificial inheritance. Therefore, artificial adaption is accomplished through producing new or modified artificial offspring otherwise known as art.

These artificial adaptations made our ancestors more evolutionary competitive by enhancing our ability to hunt animals. However, our ancestors soon repurposed these artificial adaptations for conflict within our own species. So, these art forms are also an example of ***artificial exaptation*** or the repurposing of an existing organon (i.e., instrument, tool, implement) for a new, beneficial purpose.

12.9.2   AN ARTIFICIAL ADAPTATION (ART)

An artificial adaptation is the production of a new art form or a change in an existing art form. This typically takes the form of structural or behavioral change. An artificial adaptation occurs due to a specific stimulus in the environment. It typically removes or partially removes an artificial check on homo sapiens' ability to survive or expand. For example, a change in climate can cause homo sapiens to develop a new or lose an existing artificial phenotypic trait such as a coat. However, Darwin believed that the production of new adaptations, artificial or natural, was primarily driven by interaction and/or competition – both interspecies and intraspecies.

Another example of an artificial adaptation is a projectile weapon. The slingshot is likely one of the first art forms of a projectile weapon. Since then, homo sapiens have evolved from the slingshot to the spear, bow and arrow, crossbow, rifle musket, assault rifle, etc. These artificial adaptations have provided homo sapiens evolutionary benefit that either improved our chances of survival or reproductive success.

Another term for an artificial adaptation is an ***artificial adaptive trait***. Artificial adaptive traits are beneficial as they enhance homo sapiens' ability to compete in the struggle for existence. This book will discuss two types of artificial adaptations:

Artificial Adaptation – Structural. Changes to a physical art form (i.e., a product) is an artificial structural adaptation. Artificial structural adaptations enable homo sapiens to remove constraints of our natural phenotype. Examples include cars, planes, submarines, coats, shoes, swords, guns, smartphones, etc. Utilizing these artificial adaptations homo sapiens can move faster than a cheetah, fly higher than an eagle, dive deeper than a shark, sustain arctic conditions like a polar bear, step on sharp rocks like a ram, impale threats with a rhino's horn, launch projectiles like an archerfish, and communicate over great distances like a whale's low frequency songs.

Artificial Adaptation – Behavioral. Changes to homo sapiens' instinctual behavior is an artificial behavioral adaptation. Artificial behavioral adaptations are the habits that homo sapiens develop after they are born. These are non-genetic behavioral patterns learned through experience and action. Examples include temperance (i.e., virtue) and intemperance (i.e., vice). In fact, artificial behavioral adaptations can be used to restrain natural behavioral adaptations when they are self-destructive. As this concept is a bit more complex than artificial structural adaptations, we will cover it in more detail below.

12.9.3 ARTIFICIAL INSTINCTS

"Virtue [artificial behavioral adaptations], then, being of two kinds, intellectual [habits of thought] and moral [artificial instincts], intellectual virtue [habits of thought] in the main owes both its birth and its growth to teaching (for which reason it requires experience and time), while moral virtue [artificial instincts] comes about as a result of habit, whence its name (ethike) is one formed by a slight variation form the word ethos (habit). From this it is also plain that none of the moral virtues [artificial instincts] arise in us by nature; for nothing that exists by nature can form a habit contrary to its nature."

Aristotle, "Nichomachean Ethics"

Artificial instincts are artificial behavioral patterns or habits. Artificial instincts are not inherited genetically at birth. The species homo sapiens produce artificial instincts to adapt to specific competitive and/or environmental conditions. Therefore, artificial instincts enable homo sapiens to develop an infinite number of instinctual behaviors in response to an infinite number of environmental stimuli. That is why artificial instincts vary so much across our species. Charles Darwin himself compared both natural and artificial instincts in his book *On the Origin of Species*:

"Frederick Cuvier and several of the older metaphysicians have compared instinct with habit. This comparison gives, I think, an accurate notion of the frame of mind under which an instinctive action is performed, but not necessarily of its origin."

Darwin meant this metaphorically – but it is also literal. Artificial instincts can be either developed originally or can be explicitly passed on by education. In either case, artificial instincts are forms of art. Much of our educational system's purpose is to pass on artificial adaptive traits via the process of artificial inheritance. Artificial instincts are "soft-coded" like software somewhere physically in the homo sapiens brain. Because artificial instincts are like software, they can be re-coded, if necessary, in the future. This is a truly powerful evolutionary competitive advantage.

Homo sapiens can do this because we have developed sufficient cognitive abilities to generate an internal stimulus (e.g., decisions, motivations, etc.). This provides us with the imaginative capacity to perceive both our own and other's behavioral patterns. Homo sapiens can even develop artificial instincts that can restrain our natural instincts such as our sex drive. This prevents homo sapiens from being locked into behavioral patterns, artificial or natural, that are detrimental to immediate and/or long-term survival. This provides infinite evolutionary value in the struggle for existence.

12.9.4 ARTIFICIAL CO-ADAPTATION

Artificial co-adaptation is when two or more organisms develop artificial adaptations beneficial only in the context of their interaction. We will address this concept in the section on Artificial Perfection as it is best understood in context.

12.9.5 ARTIFICIAL PHENOTYPIC PLASTICITY

"The manipular legion was flexible while the phalanx was not, and this proved the decisive factor in a clash between the two, especially at Cynoscephalae and Pydna, both of which occurred accidentally and were disorganized affairs...The Roman military system was characterized by its flexibility. The same basic structure could adapt to local conditions and defeat very different opponents."

Adrian Goldsworthy, "Roman Warfare"

The benefit of artificial phenotypic plasticity is 1) the ability to change the individual organism's artificial phenotype and 2) to do so without the process of artificial inheritance. In effect, the individual natural or artificial organism has the capacity to adjust some aspects of its artificial phenotype (i.e., structural, behavioral, etc.) without a mutation of its artificial DNA.

This artificial flexibility removes a constraint imposed by the artificial inheritance process – time. An organism, natural or artificial, can immediately initiate artificial phenotypic change in response to a direct environmental stimulus. Homo sapiens can achieve this rapid change without repeating the process of conceptual selection to produce new art. Homo sapiens simply artificially select aspects of existing art or components of art. This improves the chances of survival for both the individual organism and the species.

An example of artificial phenotypic plasticity is the Roman legion. This artificial plasticity enabled the Roman legion to defeat the previously dominant military formation – the Macedonian phalanx. The artificial plasticity of the legionary formation exposed the artificial constraints in the phalanx formation. In time the Macedonian phalanx went artificially extinct. Military organizations across the Mediterranean world adopted legionary formation as their new standard.

Another example is the flexible factory developed by Japanese manufacturers. After World War 2 Japanese manufacturers dealt with extreme competitive pressures. This forced these companies to artificially perfect their ways of working to maximize use of scarce resources. In their book *Competing Against Time,* George Stalk and Thomas Hout describe the artificial plasticity Japanese manufacturers developed to compete successfully with Western companies:

"While many Western executives worked throughout the 1970s and 1980s to reduce their costs by focusing their operations, leading Japanese manufacturers began to move to a new source of competitive advantage - the flexible factory and, later, flexible operations...The difference between traditional and flexible factories add up to competitive advantage for flexible factories in both productivity and time. The labor productivity advantage of a flexible factory can be 50 to 100 percent over a traditional factory...A strategic gap emerges between the capabilities of the flexible and traditional factories."

The ITIL 4 framework is yet another example. It is intended to provide a 21st century IT service provider the same artificial flexibility as the roman legion and the flexible factory. This is critical given the volatile change inherent in modern industries due to artificial disruption and artificial punctuated equilibrium. As stated by Axelos in ITIL 4: Digital and IT Strategy:

"Business leaders must also understand the limits of technology, and have a basic understanding of best-practice frameworks and ways of working...The latest wave of innovation had led to terms such as 'digital transformation' and 'organizational agility'...The ability to achieve and maintain position requires an organization to think differently about its business and operating models."

The Roman Republic responded to extreme evolutionary pressures by developing the legionary system. It was a key artificial adaptation that organized not only their military, but their entire society. In response to the extreme artificial evolutionary pressures of the Digital Age, organizations are developing similar patterns in artificial adaptation – specifically artificial phenotypic plasticity.

12.10 ARTIFICIAL INHERITANCE

"It developed its own culture as a diverse community, welcoming to outsiders but proud of its own, traditional way of doing things…Indeed a cosmopolitan openness to the world and fidelity to the mos maiorum, the Latin word for "ancestral custom," may have been interrelated: if social cohesion was to be maintained, the one needed to be corrected by, or balanced with, the other."

Anthony Everitt, "The Rise of Rome"

The process of artificial inheritance is about transmitting conceptual information. One or more organisms (parents) pass on information to one or more replications of itself (offspring). This transmitted information enables offspring to develop its parent's completed system of artificial adaptations. This enables the offspring to develop a variation of its parent's artificial form (i.e., structure, behaviors, etc.).

An example of structure is the clothes worn, weapons used in warfare, etc. An example of artificial instincts is social skills, fighting skills, etc. This information exchange during upbringing also enables homo sapiens to change and adapt. In the struggle for existence, a species' ability to adapt is a decisive factor. This makes the process of artificial inheritance crucial in both natural and artificial evolutionary competition.

The Romans, as usual, understood the importance of passing on artificial adaptations. They had a term for it – the "mos maiorum" meaning "way of the ancestors". The Romans understood that the patterns of existence repeat themselves. Their ancestors had over time developed artificial adaptations suited to these patterns of existence.

So, the Romans conscientiously passed on these advantageous artificial adaptations. They did so in each generation through structured education of the young. The mos maiorum organized Roman social, political, economic, and military life better than any other classical society. It was a core competitive advantage for the Roman Republic and central to their success in conquering the Mediterranean.

12.11 ARTIFICIAL GENOME

"It is evident, then, that it belongs to one science to be able to give an account of these concepts as well as of substance...and that it is the function of the philosopher to be able to investigate all things. ..But since there is one kind of thinker who is above even the natural philosopher (for nature is only one specific genus of being), the discussion of these truths will also belong to him whose inquiry is universal and deals with primary substances."

Aristotle, "Metaphysics"

Artificial genomes are at the core of artificial reproduction. The inheritance and variation of concepts is central to the process of artificial evolution. Aristotle and the ancient Greeks were the first to discover the patterns in how art was produced. They were among the first to study concepts in a scientifically rigorous manner.

In ancient Greek the word "idea" means "to see a pattern in nature". When the ancient Greeks said "nature" they meant "the world". When you conceive a new idea, you have generated a mental image that expresses a pattern. If that new idea (pattern) proves useful then it will be repeatedly conceptually selected by homo sapiens. Once a consensus on the meaning of this repeatedly used pattern is reached, it becomes a "concept". Concepts are then used exactly like genes as the building blocks for generating conceptual models for new art forms.

Conceptual information is organized in the below manner within homo sapiens' process of artificial inheritance. It is a variation of the pattern found in natural genomes. Richard Dawkins selected the metaphor of our system of books and libraries in the book *The Selfish Gene* because it is literally the same. His imagination pattern-matched a natural evolutionary pattern with an artificial evolutionary pattern without conscious awareness.

Information. Information is the artificial variation of DNA. The information contains concepts which enable us to produce new art. This was the reason the Gutenberg printing press was such an important artificial adaptation. It enabled the production of books (i.e., art) on a scale not previously possible. This increased the number of books available to homo sapiens across the world. In effect, homo sapiens had access to an ever-increasing artificial DNA code from which to produce new adaptations. The internet has taken this to a whole new level. For the first time in history we are assembling a single artificial DNA code from all homo sapiens' knowledge worldwide. The importance of this fact is stupendous.

Letters. Letters are the lowest-level building blocks of information or artificial DNA. Letters are an artificial variation of nucleobases.

Words. Words are the next-level building blocks of artificial DNA. Words are the artificial variation of nucleotides. Since artificial evolution allows for unlimited adaption our dictionaries are constantly expanding and evolving as part of artificial evolution. Our ability to produce art is only limited by the words, and most importantly concepts, we have available to our imagination.

Sentences. Sentences are the next level building blocks of artificial DNA. Sentences are the artificial variation of nucleotide sequences. As with nucleotide sequences, not all sentences contain concepts (i.e., genes). The sequence in which sentences are read affects the meaning conveyed by the paragraph or chapter.

Concepts. A concept is a specific sequence of sentences that is explicitly used to describe a known pattern in the world. Concepts are the artificial variation of genes. They are used to produce phenotypic traits (art). Like genes, concepts are combined to generate a single piece of art such as a car. Concepts are the basic unit of artificial inheritance. Each artificial phenotype has an associated conceptual model which formed the basis upon which it was generated.

Outlines. The individual concepts in an instructional manual are deliberately ordered in sequence to be read. If you change the conceptual sequence, you will begin to generate different meanings than intended by the author.

Outlines are the artificial variation of the gene sequence. Both must be read in a specific order to produce the pattern (i.e., mental, or physical) intended.

Conceptual Models. A conceptual model is the combined concepts and outline of a specific artificial DNA code. Conceptual models produce a general art form such as a car, speech, tactic, etc. Conceptual models are the artificial variation of the genotype. The designer has ordered a specific set of concepts in a specific sequence to convey a specific meaning. Genotypes work the same basic way.

Conceptual Design. Conceptual designs are the instruction manuals for producing a specific piece of art. A conceptual design is a specific expression of a conceptual model. A conceptual design takes the conceptual model and produces a specific variation of the art form. It is the adaption of the conceptual model to be fitted to the unique circumstances in which the art will be used. It is the instruction manual for producing something specific in the world. Conceptual designs are the artificial variation of chromosomes – they contain all the instructions for producing a new organism.

Artificial Mutation. Learning is the acquisition of information through experience, study, or artificial inheritance (i.e., education). It alters the artificial DNA available to individual homo sapiens and therefore the entire community. Learning is the source of all conceptual variation in thought and is, therefore, the artificial variation of mutation. Artificial mutations occur due to copying errors or in response to specific environmental stimuli (e.g., competition, climate, etc.). The mistranslation of the *Meno* by ancient sources is an example of such a copying error. Artificial mutations can occur without producing concepts or art. Artificial mutations can also alter one or more concepts. This leads to the generation of a new concept or a new artificial phenotype.

Bookcase. All the books or information sources available to individual homo sapiens, other than those already in memory, are equivalent to a bookcase. Within this bookcase is all the artificial DNA and accompanying concepts you possess to produce art. This bookcase is equivalent to the nucleus of a cell. Your imagination uses the conceptual information from your instructional manuals to produce new art.

Human Mind. The mind of individual homo sapiens is the lowest level where conceptual selection can take place. However, homo sapiens can collaborate to perform conceptual selection within a group. This is at the core of the Socratic Method described earlier. This collaboration is what Napoleon Hill called the "master mind' in his book *Think and Grow Rich*. We use the natural adaptation of imagination in combination with existing concepts to pattern-match. From this new pattern, homo sapiens produce new art forms (organon). As Napoleon hill expressed in his book – "thoughts become things and powerful things at that". In this regard, the mind of homo sapiens functions much like a cell – turning a pattern of information into the form of a physical object.

12.12 CONCEPTUAL SELECTION

"Consequently, as appears in the Topics, we must first of all have by us a selection of arguments abouts questions that may arise and are suitable for us to handle...We may now be said to have in our hands the lines of argument for various special subjects that it is useful or necessary to handle, having selected propositions suitable in various cases."

Aristotle, "Rhetoric"

Conceptual selection is a subprocess of the process of artificial selection. To produce new art, homo sapiens must conduct the process of conceptual selection. By using our imagination homo sapiens can **search** for and **select** concepts. The concepts selected are then combined to form a new pattern.

Once the process of conceptual selection is completed, a new idea, or pattern in nature, is **conceived**. This is why our ancestors selected the term "conceive" to describe this process. It is literally correct as conceptual selection is a variation of the process of sexual selection. Sexual selection ultimately leads to the conception of a new organism.

In its most basic sense, the process of conceptual selection is about pattern selection. The human mind functions like a womb by which to conceive new ideas. New ideas are often a variation of the underlying conceptual patterns found in nature. The concepts are selected from the artificial DNA (information) available to individuals or groups of homo sapiens.

Plato's reference to homo sapiens' new ideas as their "children" is literally correct. The production of art based on new ideas is equivalent to giving birth as in natural reproduction. Once a new idea reaches a state of broad consensus among homo sapiens, it is then considered a concept. This new concept is then added to the existing pool of artificial DNA (information) from which homo sapiens select.

12.13 ARTIFICIAL REPRODUCTION

"Thus, then, are all natural products produced; all other productions are called 'makings'. And all makings proceed from either art or from a faculty or from thought...from art proceeds things of which the form is in the soul [mind] of the artist...Of the productions or processes one part is called thinking and the other making...the final step of the thinking is making."

Aristotle, "Metaphysics"

Artificial reproduction is the process by which homo sapiens reproduce a mental image (pattern) conceived in homo sapiens' thought into art. The art produced is a form of artificial offspring. Artificial reproduction includes the process of conceptual selection. Homo sapiens **search** for concepts and then **select** specific concepts from available artificial DNA. We then **conceive** new ideas or patterns.

The new idea is then **developed** into a specific new design or art form using raw materials available. The design is then considered **produced** when the new art form is completed. Combined, the processes of conceptual selection and artificial reproduction follow the same general pattern:

- Homo sapiens **search** for concepts from which to generate a new conceptual model.
- Homo sapiens then **select** concepts for the new conceptual model.
- Homo sapiens then **conceive** a new conceptual model or pattern for the art.
- Homo sapiens then **develop** this conceptual model by using suitable materials.
- Homo sapiens then **produce** the new art form when it is completed.

This is the process by which all art is produced. Artificial DNA (information) is transformed into an art form in the world via this process. Evolution has perfected this process over billions of years. It is the most efficient way on earth to produce new things. Homo sapiens produces new things using a variation of the process of natural reproduction for this reason. This is how thoughts become things – information contained in our minds is transformed into art.

12.14 ARTIFICIAL HYBRIDISM

"Most of those writers who have attempted to give an authoritative description of political constitutions have distinguished three kinds, which they call kingship, aristocracy and democracy...It is clear we should regard as the best constitution one which includes elements of all three species... by which the Roman constitution was controlled...that it was impossible for the Romans themselves to declare with certainty whether the whole system was an aristocracy, a democracy or a monarchy."

Polybius, "The Rise of the Roman Empire"

Artificial hybridization is the process by which two "conceptual species" are blended to form a new concept. As a result of the process of artificial inheritance, the new concept produced will provide value to human activity or not. If the concept is useful, it is likely to be conceptually selected. New art typically follows as homo sapiens artificially adapt the art using the new concept to fit their specific needs. If the concept is not useful it will not be conceptually selected and will likely disappear from human activity.

Artificial hybridization is an extension of evolution's strategy of variation. The entirely new hybrid concept combines two different artificial genomic lines or patterns. As a result, a new evolutionary trajectory is created for the hybrid concept distinct from those of the parent concepts.

This offers new possibilities for artificial adaptation that were previously unavailable in the parent concepts. Artificial adaptations generally facilitate the removal of constraints on homo sapiens expansion in its ecosystem. This increases homo sapiens' probability of survival in the struggle for existence.

The Roman constitution is a great example of this conceptual hybridization. The Romans evolved this constitution through artificial adaptations in response to specific political stimulus. This led to the blending of three different forms of government (i.e., monarchy, aristocracy, and democracy) which created a relative artificial equilibrium within the Roman State. This allowed the Romans to mostly eliminate civil strife and civil wars for almost five centuries. This was a distinct competitive advantage for the Romans which Polybius points out as central to Rome's imperial expansion.

Another more modern example is agile project management. This project management approach was produced by the conceptual blending of traditional project management techniques and lean thinking. It produced a new hybrid concept of agile project management. This produced a new project management approach which is indispensable in the evolutionary conditions of the Digital Age.

12.15 ARTIFICIAL VARIATION

"In the late 1970s, Japanese companies exploited the benefits of flexible manufacturing to the point that a new competitive thrust emerged - a variety war...the battle that erupted between Honda an Yamaha for supremacy in the motorcycle industry...Honda used expanding variety to bury Yamaha under a flood of new products...Demand for Yamaha motorcycles dried up...so decisive was its victory that Honda effectively had as much time as it wanted to recover...Variety had won the war."

George Stalk & Thomas Hout, "Competing Against Time"

The process of artificial reproduction involves the searching and selecting of concepts. These concepts are then blended to intentionally conceive a new pattern of art. However, the new conceptual pattern is not always a dramatically different one. Often the new conceptual pattern is a slight modification of the original pattern. This produces an artificial variation of the original art form.

Examples of artificial variations are:

- The different motorcycle product lines of Honda, Yamaha, and Harley-Davidson.

- German blitzkrieg tactics vs. American combined arms tactics in World War 2.
- The liberal and conservative subcultures of the United States.
- The American and National Leagues in American baseball.
- Private and public universities in the United States.

Artificial variations of art are everywhere. Often the variations are produced to address the specific natural ecosystem conditions in which homo sapiens operate. The cuisine of different cultures or artificial species often vary. This artificial variation is typically produced via adaption to local circumstances (i.e., climate, foods available, cultural preferences, etc.).

Rice grows well in China so that culture bases its cuisine on rice. Wheat grows well in Italy so that culture bases its cuisine on pasta. They are both foundational carbohydrate foods. You see this same artificial variation in Africa with the carbohydrate cuisine base of nshima. Homo sapiens work with the "clay" they have which then produces artificial variations.

12.16 ARTIFICIAL FITNESS

"Though the principles of banking trade...is capable of being reduced to strict rules. To depart upon any occasion from those rules...is almost always extremely dangerous, and frequently fatal, to the banking company which attempts it. But the constitution of joint stock companies renders them in general more tenacious of established rules than any private copartnery. Such companies, therefore, seem extremely well fitted for this trade."

Adam Smith, "The Wealth of Nations"

Artificial adaptations or art are a means to specific ends for homo sapiens – survival, reproduction, pleasure, etc. Therefore, the artificial fitness of artificial adaptations is directly tied to their value in achieving those ends. Artificial adaptations are produced by selecting concepts for their design. The more valuable homo sapiens find a concept, the more it will be conceptually selected and expressed in art forms.

An example of this is lean thinking. Lean thinking was originally conceptually selected for the art form of automobile manufacturing. Lean Thinking was then conceptually selected and adapted for the art form of agile project management. This created an artificial variation of the lean thinking art form which generally included the same conceptual pattern.

The value of lean thinking has caused its rapid expansion across human activity. This causes lean thinking to have a high degree of artificial fitness. The conceptual pattern of lean thinking is being artificially reproduced in various art forms each day.

Cultures or artificial species are a slight variation of the concept of artificial fitness. Culture is thought to have artificial fitness when it spreads to large homo sapiens populations. In this case the pairing is one-for-one individual homo sapiens who possesses sufficient artificial adaptations of the culture. The more a culture spreads to individual humans the higher its fitness level.

An example is ancient Roman culture. In 753 BC the Roman population and its cultural influence on its neighbors was very small. Thus, the culture had low artificial fitness. But by 100 AD the Romans had conquered the Mediterranean and spread their culture across many regions, ethnicities, etc. At this time the Roman culture had high fitness relative to other cultures on the planet.

However, in 1453 AD when the Ottoman Empire conquered Constantinople (Istanbul), it ended the Byzantine Empire. This empire was the last cultural stronghold of the original Roman culture. Once Constantinople was conquered, the ancient Roman culture effectively went extinct. Its artificial fitness ceased to exist, although many artificial adaptations of the Roman culture still live on in modern Italian culture.

12.17 ARTIFICIAL SELECTION

"Then there must be a selection [artificial selection]. Let us note among the guardians those who in their whole life show the greatest eagerness to do what is for the good of their country, and the greatest repugnance to do what is against their interests...Those are the right men."

Plato, "The Republic"

Artificial selection is the process by which new and existing art is intentionally selected by homo sapiens for use in human activity. The intentionality of artificial selection requires conscious thought to perform. This is why on earth only homo sapiens (at least we think) can perform artificial selection. It is what distinctly separates us from every other natural species. It is why we are the keystone natural species on the planet.

During artificial selection homo sapiens select artificial adaptations or art for use in either artificial (e.g., business, sports, etc.) and/or natural (e.g., warfare, mating, etc.) competitive contexts. The art is selected for its evolutionary value for survival, mating, pleasure, etc. Artificial selection is performed by both individuals and groups of homos sapiens. If you are reading this sentence, you likely artificially selected this book (i.e., art) when you purchased it.

Artificial selection functions exactly the opposite of natural selection. Art is not self-sorted randomly but is sorted with intentionality by homo sapiens. Over time homo sapiens intentionally produce new art forms. The most beneficial art is then artificially selected by homo sapiens and reproduced in increasing quantities. Often the new art is an improvement or enhancement of existing art. This is accomplished via the process of conceptual selection. This effectively evolves the art form along an adaptive trajectory. This is a variation of the process of natural evolution. But isn't natural, it is artificial.

Interestingly, just as in natural evolution, variations of similar patterns are artificially selected by homo sapiens again and again. An example is projectile weapons. Beginning with the sling homo sapiens have incrementally evolved the concept and art of projectile weapons. This has led us to assault rifles, cruise missiles, and nuclear missiles that essentially perform the same basic function in warfare. However, this function is performed on a scale and with a severity unimaginable when first invented. As once again we return to Charles Darwin's profound insight from his book *On the Origin of Species*:

"We can, in short, see why nature is prodigal in variety, though niggard in innovation. But why this should be a law of nature if each species has been independently created no man can explain...Many other facts are, as it seem to me, explicable on this theory...these facts cease to be strange, or might even have been anticipated."

Why is art produced by homo sapiens so essentially "prodigal in variety, yet niggard in innovation"? This is due to the same reasons the pattern of a crab is consistently advantageous in natural evolution. The environmental and competitive patterns in both natural and artificial evolution repeat.

Otherwise, we wouldn't have evolved projectile weapons from the sling to the assault rifle so predictably. As the patterns of natural law and evolution continue to repeat, variations of the same phenotypic patterns will tend to be selected – both naturally and artificially.

Below are some examples of artificial selection familiar to us all:

Selective Breeding. We covered this example in detail in chapter 11. Please refer to that content.

Voting. Homo sapiens perform the act of voting in democratic systems of government. The art form being selected are political candidates. A candidate is a combination of naturally inherited and artificially adapted phenotypic traits. However, together the pattern of a candidate and his or her specific policy platform are essentially a form of art. Voting for a candidate is the act of artificial selection.

Purchasing. Homo sapiens perform the act of purchasing in economic systems. Things purchased are generally products or services. Both are art forms. Products are artificial structural adaptations. Services are typically a combination of artificial adaptations (the artificial objects and instincts) and natural adaptations (the people and their natural instincts). When you make a purchase from a business you artificially select the artificial variation they produced.

Dressing. Homo sapiens perform the act of dressing each day. Dressing involves the act of artificial selection. Each piece of clothing is art and is artificially selected to be worn. However, dressing also involves conceptual selection. The selection of an entire outfit to produce a specific look is a form of self-expression or art. The total outfit pattern is conceptually selected and then each piece of clothing is artificially selected to fit that pattern. This is an example of where the two forms of selection work together seamlessly in human activity.

12.18 ARTIFICIAL COEVOLUTION

"Rome's military greatness was essentially built on the power of her legionary infantry...only in Scipio's brief passage across the stage do we find a real break with this tradition...From his arrival in Sicily onwards Scipio bent his energies to developing a superior cavalry, and Zama, where Hannibal's decisive weapon was turned against himself, is Scipio's justification."

B.H. Liddell Hart, *"Scipio Africanus: Greater Than Napoleon"*

The relations of natural organisms are the single most important factor in natural evolution. This is due to scarce resources available in the struggle for existence. This resource constraint is a crucial driver of conflict between homo sapiens. We compete at the individual and group levels for food, territory, and to reproduce. As a result, homo sapiens continuously artificially adapt to maximize the value derived from scarce resources.

Homo sapiens' competition with other species was originally the prime driver of artificial evolution. We coevolved in competition with other species such as saber tooth tigers and wholly mammoths. Then at some point we artificially evolved to become the keystone species in natural ecosystems. This likely occurred after we displaced and absorbed our genetic cousins such as Neanderthals from their territories. Since that point in time the prime driver of artificial evolution has been intraspecies competition.

Most artificial evolution now occurs in intraspecies competition such as warfare, politics, business, sports, academia, mating, etc. Warfare traditionally occurred in natural ecosystems. But with the advent of cities warfare began to occur in artificial ecosystems. In addition, with the invention of cyberspace warfare now happens in the cloud and wide area networks. All other intraspecies competition takes place in artificial ecosystems (e.g., sports arenas, industries, nations, societies, etc.).

Artificial evolution typically occurs when two or more homo sapiens, artificial organisms, or artificial species have sustained interaction. This creates a persistent selective pressure on all entities involved. This selective pressure then directly affects the artificial evolutionary trajectory of all.

Coevolution often occurs as an ever escalating 'arms race' between parties. The competition between the artificial species of ancient Rome and Carthage is a great example. In the First Punic War Carthage was a dominant sea power. Rome was a dominant land power without any naval tradition. Once the war started Rome artificially evolved into a dominant sea power. Rome then defeated Carthage at sea, winning the war. Rome then repeated the process to win the Second Punic War. They did so by copying another Carthaginian artificial adaptation – cavalry.

Coevolution often leads to unintended consequences in homo sapiens competition. If you had told the Romans fighting in the First Punic War victory would lead to the loss of their political liberty, they likely would have reconsidered their options. The average Roman citizen was fighting to defend the existence of the Republic itself. This is what safeguarded their liberty, their property, and their lives.

But once a new artificial evolutionary trajectory begins it often takes on a life of its own. This can be different from what any of the parties that initiated the competition intended. This holds true in war, politics, business, sports, etc. This is the concern Geoffrey Hinton has regarding the competition between Microsoft and Google to produce increasingly powerful variations of AI.

12.19 ARTIFICIAL CO-ADAPTATION

"Service...A means of enabling value co-creation by facilitating outcomes customers want to achieve...digital leaders need to...increase collaboration and partnering with consumers to enable value co-creation. There cannot continue to be siloed work or a large separation between IT and the business."

Axelos, "ITIL 4: Digital and IT Strategy"

Artificial co-adaptation is when two or more homo sapiens, artificial organisms, or artificial species develop new interdependent adaptations. The adaptations are beneficial only in the context of their interaction. In effect, the entities involved adapt their individual patterns (i.e., structure, behavior, etc.) to integrate with each other. This creates evolutionary value otherwise not available to the entities. This increases the involved entities' chances of being naturally and/or artificially selected in the struggle for existence.

An example of this is the United States Air Force's (USAF) Enterprise IT-as-a-Service (EITaaS) solution. The USAF has outsourced its basic IT services to commercial vendors. Part of the intent of the EITaaS program is to develop co-adaptive relationships with the commercial vendors. In effect, a major objective of the USAF for EITaaS is to leverage the value of evolutionary co-adaptation.

As part of the relationship, both the USAF and the vendors will co-create value as described within ITIL 4. Co-creating value is basically intentionally entering into a co-adaptive artificial evolutionary trajectory with stakeholders. It is a central concept of the ITIL 4 Framework. This will drive the artificial evolution of all parties involved.

The EITaaS solution is explicitly implementing the ITIL 4 framework for managing and delivering its services to the enterprise. Therefore, the EITaaS program will create co-adaptive relationships at all levels of execution. This creates a win-win scenario for all stakeholders long-term. The USAF has wisely chosen an effective evolutionary strategy to achieve its strategic objectives. If evolutionary history is any indicator, then the USAF will gain increasing competitive value from the EITaaS solution over time.

12.20 ARTIFICIAL PERFECTION

"Soc. ...and whatever is the shuttle best adapted to each kind of work, that ought to be the form which the maker produces in each case....And the same holds true of other instruments: when a man has discovered the instrument which is naturally adapted to each work, he must express this natural form, and not others which he fancies, in the material, whatever it might be, which he employs..."

Plato, "Cratylus"

Plato perfectly described artificial perfection in his dialectic *Cratylus*. Plato states that the organon or instrument is perfectly adapted when the means cannot be improved to achieve a specific end. Products of art are no different than products of nature in this respect. Art is perfected when any additional investment in that instrument would yield no additional benefit and only create waste. This is succinctly expressed by Adam Smith in his book *Wealth of Nations*:

"And thus the certainty of being able to exchange all that surplus part of the produce of his own labour, which is over and above his own consumption, for such parts of the produce of other men's labour as he may have occasion for, encourages every man to apply himself to a particular occupation and to cultivate and bring to perfection whatever talent or genius he may possess for that particular species of business."

Plato and Adam Smith are expressing the same evolutionary concept. This is the same evolutionary concept found in species in natural ecosystems. Natural evolution eventually mercilessly punishes a wasteful phenotype. Of course, there is no absolute perfection in artificial evolution – only a form of relative artificial perfection. But artificial evolution moves with such velocity and diversity that artificial perfection for most organon, or instruments, doesn't last very long. The authors of *Lean Thinking* described this challenge directly:

"Paradoxically, no picture of perfection can be perfect. If the value stream for automotive glass could be reconfigured as we suggest, it would then be time (immediately!) to imagine a new perfection which goes even further. Perfection is like infinity. Trying to envision it (and to get there) is actually impossible, but the effort to do so provides inspiration and direction essential to making progress along the path."

That's right, Lean Thinking is the artificial variation of natural perfection. That means the art being perfected by lean methodologies may change infinitely, but lean thinking never will. It is one of the pure forms for which Plato searched.

When homo sapiens produce new art the first version is often wasteful in some regard. Over time in interspecies competition the original waste is forced out of the art. Lean Thinking is simply the proactive version of this competitive process which is inherently reactive.

Lean Thinking as a concept is itself already perfected. It was perfected by natural evolution over billions of years. It wasn't invented, it was a discovered pattern in nature. This is why it works and will eternally work in human activity.

12.22 ARTIFICIAL EQUILIBRIUM

"The notion of an equilibrium point is the basic ingredient in our theory. This notion yields a generalization of the concept of the solution of a two-person zero-sum game…In the immediately following sections we shall define equilibrium points and prove that a finite non-cooperative game always has at least one equilibrium point."

John Nash, "Non-Cooperative Games"

Equilibrium in artificial ecosystems occurs because of the imaginative capacity of homo sapiens. Homo sapiens operate in a persistent set of competitive and environmental conditions – some natural and others artificial. Over time a pattern emerges within homo sapiens behavior within the artificial ecosystem. This does not mean that the artificial ecosystem loses its inherently dynamic nature. It only means that new changes happen within the same persistent patterns.

This is the variation of natural equilibrium – artificial equilibrium. Otherwise known as the "Nash equilibrium". It is homo sapiens' variation of natural equilibrium given our capacity to imagine future scenarios. Cooperative games operate based on the same concept except in reverse. Competitors co-adapt in a similar manner (price-fix) to establish the equilibrium.

When homo sapiens can no longer imagine any additional benefit of artificial adaption they will cease artificially evolving. This is because we tend not to incur an additional cost without any accompanying benefit. This is natural perfection being enacted via our natural instincts. It halts artificial evolution from progressing down a specific evolutionary trajectory.

Homo sapiens may incur short-term costs for perceived long-term benefit. But the benefits must increasingly outweigh the immediate cost over an increasing time horizon. Irrational and/or unintelligent homo sapiens may incur costs for perceived immediate benefits which are not realistic. This scenario is only possible due to the consciousness of homo sapiens.

Artificial equilibrium creates relative stability in the artificial ecosystem in terms of competitive and environmental dynamics. This relative stability exists for a protracted period. When this relative equilibrium is reached predictable patterns will recur over time.

These patterns will recur in response to recurring events such as economic downturns, election results, minor military conflicts, etc. All the artificial species/organisms will become artificially perfected within this overarching set of persistent patterns. This relative equilibrium will persist in an artificial ecosystem until a disruption occurs.

A great example of relative artificial equilibrium is the Cold War between the United States and the Soviet Union. Nuclear weapons placed a military constraint on both nations. Therefore, a large-scale conventional war such as World War 2 was not possible.

As a result, both nations turned to competition of another kind. Both nations perfected their diplomatic, intelligence, and economic capabilities in a new form of intraspecies competition – a "Cold War". The Cold War's relative artificial equilibrium ended with the fall of the Soviet Union. Then a new relative artificial equilibrium emerged worldwide.

12.23 ARTIFICIAL DISRUPTION

"But inventions alone don't create huge value. Not unless they disrupt something...Recast in terms of disruptive innovation...reveals two facts about historical disruption that have always been true: 1) Those disruptions do best that are aimed at a core benefit the end user understands; and 2) For most of history, disruptions occurred in a physical world...Digital disruption will change that."

James McQuivey, "Digital Disruption"

Artificial disruptions are events that permanently alter one or more persistent patterns of an artificial ecosystem. The source of the disruptive event can be either internal, external or both. Disruptive events typically occur when an existing constraint is either removed or becomes more restrictive.

An example is when an artificial species or organism within an artificial ecosystem develops a new artificial adaptation. This new artificial adaptation enables the artificial species or organism to enhance access to an existing resource (i.e., territory, food, consumers, information, wins, etc.).

A period of turbulence will then occur in the artificial ecosystem as an altered dynamic of competition arises. The artificial ecosystem will then settle into a new form of relative artificial equilibrium. This relative artificial equilibrium will hold until the next disruptive event.

A practical example of artificial disruption is the NFL's 21st century rule changes. The NFL has artificially imposed constraints on restricting receiver movement and violence against the quarterback. Each rule change was a disruptive event. The NFL teams then artificially adapted by changing their strategies – draft (e.g., smaller receivers, edge rushers), offensive (e.g., mobile quarterbacks, increased passing), defensive (e.g., generate turnovers, limit chunk plays), etc.

A second example is electric cars. Companies such as Tesla entered the automobile industry and captured market share. However, the existing automobile manufacturers adapted. They produced both electric and hybrid vehicles in response. The electric car changed the industry, but it wasn't an extinction event for most companies in the industry. Instead, the automobile industry has settled into new recurring patterns of competition that include electric vehicles.

A third example is the newspaper industry. The advent of social media (e.g., Facebook) disrupted the newspaper industry. Consumers increasingly receive their news from social media applications such as Facebook. In response existing print media (e.g., Washington Post) produced an online presence and mobile applications in response to the threat. Print media evolved to meet the competitive challenges of the Digital Age. The print media industry has now settled into new recurring patterns that now includes competition with social media companies.

12.22 ARTIFICIAL PUNCTUATED EQUILIBRIUM

"Darwinian evolution is a force of continuous change...By contrast, punctuated equilibrium suggests that evolution occurs as a series of bursts of evolutionary change...Today we are seeing a burst of evolutionary

change – *a mass extinction among corporations and a mass speciation of new kinds of companies.*"

Thomas Siebel, "Digital Transformation"

We have already covered Thomas Siebel's concept of artificial punctuated equilibrium. He got this chain in the logical pattern of artificial evolution spot on. There is nothing to add.

12.23 "THE ONLY GAME IN TOWN"

"We meet with this admission in the writings of almost every experienced naturalist; or as Milne Edwards has well expressed it, Nature is prodigal in variety, but niggard in innovation. Why, on the theory of Creation, should there be so much variety and so little real novelty?"

Charles Darwin, "On the Origin of Species"

What is a submarine if not an artificial whale? What is a plane if not an artificial bird? What is a car if not an artificial horse? What is a sword if not an artificial rhinoceros' horn? What is cell phone communication if not a whale's song? We have been artificially reproducing and describing variations of the natural world since artificial evolution became possible. There is nothing else to reproduce or describe.

Evolutionary history, both artificial and natural, is a repetition of the same patterns in varied forms. This means everything see you today, including art produced by homo sapiens, is a variation of past evolutionary realities. Homo sapiens' art must be a variation. The "One" pattern has repeated for billions of years. On earth the "One" single pattern is – *__the only game in town__*. When homo sapiens gained the natural adaptation of imagination it created a new variation of the pattern of natural evolution:

Artificial Evolution

13

CHAPTER 13: THE THEORY OF ARTIFICIAL EVOLUTION

"But if on the other hand art imitates nature, and it is the part of the same discipline to know the form and the matter up to a point (e.g. the doctor has knowledge of health…and the builder both of the form of the house and of the matter, namely that it is bricks and beams, and so forth): if this is so, it would be the part of physics also to know nature in both its senses."

Aristotle, "Physics"

13.1   ARTIFICIAL EVOLUTIONARY THEORISTS

13.1.1   ARISTOTLE

"For things different in kind are, we think, completed by different things (we see this to be true both of natural objects and things produced by art, e.g. animals, trees, a painting, a sculpture, a house, an implement)…"

Aristotle, "Nichomachean Ethics"

Aristotle was both a natural and artificial evolutionary theorist. He truly was a polymath. The word genius does not seem to do the man justice. Aristotle theorized about both forms of evolution combined in many of his writings. He saw it all as "nature". In his work *Nichomachean Ethics* is the secret to unlocking human happiness via new breakthroughs in the fields of psychology and sociology. Charles Darwin was indeed wise to choose such a mentor to shape his thinking.

13.1.2 PLATO

"Why, I said, the principle has been already laid down that the best of either sex should be united with the best as often, and the inferior with the inferior, as seldom as possible...if the flock is to be maintained in first-rate condition. Now these goings on must be a secret which the rulers only know, or there will be further danger of our herd, as the guardians may be termed, breaking out into rebellion."

Plato, "The Republic"

Plato was an artificial evolutionary theorist. Unlike Aristotle, he focused on artificial evolution. This is because Plato was an evolutionary theorist that specialized in one species – homo sapiens. In the *Republic,* he explicitly endorses selective breeding for producing genetically superior homo sapiens. The rest of Plato's work explicitly centers on artificial evolutionary theory. In addition, his "Theory of Forms" is really a theory of evolutionary patterns. He was entirely focused on unconcealing *"**the only game in town**."*

13.1.3 THOMAS SIEBEL

"Darwinian evolution is a force of continuous change...By contrast, punctuated equilibrium suggests that evolution occurs as a series of bursts of evolutionary change...Today we are seeing a burst of evolutionary change – a mass extinction among corporations and a mass speciation of new kinds of companies."

Thomas Siebel, "Digital Transformation"

The theory of artificial punctuated equilibrium has repeatedly been proven valid throughout history. The automobile rapidly replaced the horse and carriage industry. Many governments (e.g., Russia, Germany, Italy, Austria-Hungarian, etc.) collapsed after World War 1. Donald Trump's 2016 Presidential Campaign dramatically changed the Republican Party. Each is an example of artificial punctuated equilibrium.

Thomas Siebel is literally an artificial evolutionary theorist. He successfully pattern-matched the artificial variation of a natural evolutionary pattern. His book *Digital Transformation* is a work of both technology and science.

13.1.4 JOHN NASH

"The notion of an equilibrium point is the basic ingredient in our theory. This notion yields a generalization of the concept of the solution of a two-person zero-sum game...In the immediately following sections we shall define equilibrium points and prove that a finite non-cooperative game always has at least one equilibrium point."

John Nash, "Non-Cooperative Games"

Professor John Nash was an artificial evolutionary theorist. His non-cooperative games theory of equilibrium is literally artificial evolutionary theory. The cooperative game theory completes the other half of the artificial equilibrium concept. His theory was even more meaningful to homo sapiens than he knew. It was his superior imaginative capacity that enabled him to pattern-match a concept in artificial evolution. Professor Nash truly did have a "beautiful mind".

13.2 THE THEORY OF ARTIFICIAL EVOLUTION

"We meet with this admission in the writings of almost every experienced naturalist; or as Milne Edwards has well expressed it, Nature is prodigal in variety, but niggard in innovation. Why, on the theory of Creation, should there be so much variety and so little real novelty?"

Charles Darwin, "On the Origin of Species"

The theory of artificial evolution is based on the statement of Charles Darwin immediately above. It was clear that artificial selection did not really fit the pattern of natural evolution. Artificial selection is the key mechanism of an entirely different conceptual pattern. This conceptual pattern is a variation of natural evolution.

The theory of artificial evolution reveals an underlying pattern in human activity. The key mechanism of this underlying pattern is artificial selection. This theory brings clarity to our species' understanding of our artificially created reality. It allows us to understand how the species homo sapiens has artificially adapted, evolved, speciated, and expanded across the planet earth. It also helps to explain how a new variation of artificial speciation represents an extinction risk to homo sapiens. The theory of artificial evolution can generally be expressed in the following simple statements:

- Homo sapiens naturally evolved to produce the natural adaptation of imagination.
- Imagination enabled homo sapiens to intentionally perform artificial and conceptual selection.
- Conceptual/artificial selection enabled homo sapiens to produce art forms.
- Art forms enable homo sapiens to artificially vary our structure and behaviors.
- Through artificial inheritance we passed on artificial genomes and adaptations to our offspring.
- The result is that homo sapiens' population size has grown from thousands to billions.
- Artificial evolution has enabled homo sapiens to become the planet's keystone natural species.
- Homo sapiens' population size and imaginative capacity dramatically lower extinction risk.
- The result is that homo sapiens continues to be naturally selected by the process of natural evolution.

Artificial evolution has only one real constraint – perceptive. It is only a poverty of imagination that restricts the evolutionary trajectory of art. Homo sapiens have unlimited options for artificial adaption. Unlike every other species on earth we have few, if any, constraints on our artificial evolutionary possibilities. Homo sapiens can evolve an "elephant" into a "mouse". Think of watching TV in 1950 on a huge device and watching the exact same program on your smartphone today.

With imagination we can select an infinite number of trajectories for our artificial structural and behavioral development. This is what artificial evolution is – the infinite flexibility to gradually progress along any artificial evolutionary trajectory we can imagine. By so doing, homo sapiens ensure we always get naturally selected in the process of natural evolution.

That's it – the power of the theory of artificial evolution lies in its simplicity. What could be simpler than imagination? Imagination is the adaptive trait that dramatically increases our species' chance of achieving the "big win" – to never go extinct. Very few natural species have avoided extinction. Scientists think over 5 billion species have existed during earth's history. Over 99% of all those species have gone extinct. So, very few natural species have achieved the "big win" in earth's evolutionary history.

Artificial evolution is happening every moment of every day everywhere on earth. It has so been since the first homo sapiens born with imaginative capacity began producing art. This pattern will continue to repeat for as long as there are organisms on earth capable of artificial and conceptual selection.

However, the species doesn't necessarily have to be homo sapiens. Any species capable of artificial selection and conceptual selection would automatically become a threat to homo sapiens. Any such species would eventually come into competition with homo sapiens in the struggle for existence. But before we address that subject, we must first delve into homo sapiens' ultimate adaptation – imagination.

14

CHAPTER 14: THE ULTIMATE
ADAPTATION – IMAGINATION

"Gradually, through the process of natural selection, the brain developed increasingly sophisticated circuitry. And as that circuitry grew, it spawned something that Einstein said was more important than knowledge: imagination."

Joseph Jebelli, "How the Mind Changed"

14.1   ANCIENT GREECE

14.1.1   MENO

"<u>SOCRATES:</u> So, Meno, our argument has led us to suppose that the excellence [imagination] of good people comes to them as a dispensation awarded by the gods."

Plato, "Meno"
Robin Waterfield, "Plato: Meno and Other Dialogues"

In *Meno*, Plato describes the greatest homo sapiens excellence as imagination. Imagination is the natural adaptation that enabled homo sapiens to adapt and become perfected in any environment. It is homo sapiens' most important evolutionary advantage. It is made possible by human consciousness, pattern-matching, and imagination.

Plato concludes in *Meno* that something as powerful as imagination could only be a gift of the gods. But how did the gods gift imagination to homo sapiens? The answer lies in what imagination is in practical terms. Imagination is the light in the darkness for our civilization. What gift of the gods brought light to homo sapiens civilization – the fire Prometheus stole from Mount Olympus.

14.1.2 PROMETHEUS

"Fire is the key that opens all these doors and lays the foundation of human life. Without it, there is no possibility of advancement or civilization. With it, and with Promethean intelligence, who knew whether men might not become as gods themselves?...Prometheus gave fire, life, and civilization to mankind, and it could not be taken back...They were safe now; they would survive, and even, in ages to come, make themselves the dominant species on the broad face of the earth."

Robin Waterfield, "The Greek Myths"

Prometheus gave fire to man. Fire is the ancient Greek symbol for the natural adaptation of imagination. This truth is hidden in the quote above (from the book *The Greek Myths*). It was not fire that enabled homo sapiens to become the dominant species on earth, but imagination.

In the context of primitive societies, fire makes sense as a stand-in for imagination. The use of fire for cooking caused the homo sapiens brain size to grow dramatically. This then led to rapid advancements in knowledge, technology, and civilization.

In ancient Greek myth these advancements are all attributed to the fire Prometheus gave to homo sapiens. These two adaptations, one artificial (fire) and one natural (imagination), really go hand-in-hand in homo sapiens evolution. Imagination enabled homo sapiens to unconceal the secrets of nature (the gods).

14.1.3 THE SYMPOSIUM

"Those who are pregnant in the body only, betake themselves to women and beget children – this is the character of their love...But souls [mind] which are pregnant – for there certainly are men who are more creative in their souls [mind] than in their bodies – conceive that which is proper for the soul [mind] to conceive or contain. And what are these conceptions? – wisdom and virtue in general. And such creators are poets and all artists who are deserving of the name inventor."

Plato, *"Symposium"*

The ancient Greeks understood reproduction in both its forms – natural and artificial. Plato's dialogue *Symposium* directly expressed this understanding. Wise mortals harness the power of desire to conceive new ideas. These ideas are then transformed into new art or inventions. The god of desire in ancient Greek mythology was Eros.

14.1.4 EROS & PSYCHE

"...Apuleius also drew upon the Platonic texts...the idea of Love as a daemon, leading the Soul [mind] up to union with the divine, in the Symposium...inspire Psyche's clinging to Cupid [Eros]...save her life and render her [Psyche] immortal...This was the conception of Cupid [Eros] as a pure, celestial love which ennobles the soul [mind], but also as the impulse to sexual desire and procreation."

Jane Kingsley-Smith *"Cupid: Early Modern Literature & Culture"*

The Roman depiction of Eros as a small, chubby boy is a Christianization of the deity. In ancient Greek myth Eros is a virile young man. He embodied desire and profound artistry. The ancient Greeks recognized Eros as the force by which homo sapiens produced new forms – both natural and artificial.

Eros married a mortal woman named Psyche. Her name means "soul" or "mind" in ancient Greek. This makes sense as the Greeks saw the mind as a "womb" where ideas were conceived. So, desire could provide the energy to fuel our imagination.

We search for new information, new and existing, that we use to "impregnate the mind". Then with our imagination, powered by desire, we perceive a pattern in this information. We then select the pattern to form a new mental image. The result is the conception of a new pattern in nature – an idea. This fits the pattern of how artificial reproduction works in artificial evolution. The ancient Greeks perceived this pattern in their myths.

In the myth Psyche only becomes a goddess after passing many trials. This is a metaphor for human activity. Simply possessing imagination was not enough – mortals must possess sufficient desire to persevere. Only through such perseverance can mortals unconceal the secrets of the gods. An individual that possessed the combination of desire and imagination capable of this feat was truly godlike. To achieve this godlike state was the goal of the ancient Greek philosophers.

14.2 THE 20TH CENTURY

14.2.1 NAPOLEON HILL

"Truly, "thoughts are things", and powerful things at that, when mixed with purpose, persistence and a <u>burning desire</u> for their translation into riches or other material objects."

Napoleon Hill, "Think and Grow Rich"

The above excerpt is from the book *Think and Grow Rich,* published by Napoleon Hill. His book was published in 1937, a time of dramatic technological change. It has sold over 15 million copies since it was published. The book is a practical framework for performing the process of artificial reproduction. Fundamentally the book's objective is to instruct homo sapiens in how to turn "thoughts" (a pattern) into "things" (art).

The above excerpt is the same pattern present in the myth of Eros and Psyche. Both patterns possess the same underlying concepts of desire, the mind, imagination, persistence, artistry, etc. Napoleon Hill independently discovered the same process of artificial reproduction the ancient Greeks did. They all knew that the key elements of this process were found in homo sapiens – desire and imagination.

In chapter 6 "Imagination: The Workshop of the Mind", Napoleon Hill also describes the process of conceptual selection. He divides imagination into two variations – synthetic and creative. Creative imagination is the process of "inspiration" by which new ideas or future concepts are conceived. This is the gift of the gods Plato describes in the Meno. Synthetic imagination is the process of combining existing concepts into new patterns. Together both can be used to produce new art via the process of artificial reproduction.

Napoleon Hill is an artificial evolutionary scientist. He discovered the processes of artificial reproduction and conceptual selection. The express aim of the book is producing art that will be artificially selected. Napoleon Hill's framework is spot on. He got it right to a degree of precision that is quite amazing.

14.2.2 ALBERT EINSTEIN

"I believe in intuitions and inspirations...I was not surprised when the eclipse of May 29, 1919, confirmed my intuitions. I would have been surprised if I had been wrong...I am enough of an artist to draw freely upon my imagination. Imagination is more important than knowledge. Knowledge is limited. Imagination encircles the world."

Albert Einstein, "Interview: What Life Means to Einstein"

Professor Albert Einstein is one of the greatest scientists in history. He discovered a new pattern in nature – the theory of relativity. Einstein's theory was so powerful that it reshaped both the scientific understanding of nature and the field of modern physics. The theory of relativity is considered one of the most famous mathematical equations in history. In 1921, Einstein received the Nobel Peace Prize.

Einstein explicitly states it was his imaginative faculty that enabled him to make this scientific discovery. In effect, Einstein formed a True Belief about a pattern in nature which has since been chained to knowledge by the field of physics. It is exactly the process that Plato described in the *Meno*. This is why Einstein thought that imagination was more important than knowledge – imagination creates knowledge.

Einstein believed imagination was supreme. This conflicts with the supremacy of empiricism and the scientific method. Since the dawn of the Scientific Revolution in 1543, empirical fact has reigned supreme. This Scientific Revolution was an abandonment of ancient Greek thinking – a process of thinking described in *Meno* for working with evolutionary concepts. Unknowingly, Einstein is espousing a return to what we now know ancient Greek thinking really was – and he was right.

14.3 THE THEORY OF IMAGINATION

14.3.1 THE ADAPTATION OF IMAGINATION

"In March 2016 researchers saw a glimpse of imagination in action… this brain area is known to act as a hub connecting different parts of the brain…we constantly adapt to an environment that we have also shaped, gaining new knowledge while simultaneously learning from the past and imagining the future. We go beyond 'what is' to 'what could be'. We ceaselessly create new ideas…"

Joseph Jebelli, "How the Mind Changed"

Neuroscience is still in the process of discovering exactly what imagination is and how it works. However, we know that it is a natural adaptation of some kind. Therefore, like all natural adaptations it has a specific pattern. Part of that pattern is a network of connections to different parts of the brain. It seems that imagination draws on the diverse adaptations of the brain to conceive new patterns. So, imagination is a pattern used to conceive new patterns. Imagination does this by forming mental images. Nicola Tesla described the practical use of imagination succinctly:

"When I get a new idea, I start at once building it up in my imagination, and make improvements and operate the device in my mind. When I have gone so far as to embody everything in my invention, every possible improvement I can think of, and when I see no fault anywhere, I put into concrete form the final product of my brain."

Tesla described the process of designing new inventions (art). He pictured the image in as much detail as he could before producing the art form. Consider that he conducted this process over time. Therefore, he had to store the uncompleted image of the art somewhere in the brain. This means the image effectively exists in the universe somewhere.

The information that comprises the mental image is stored in at least one cell somewhere in the brain. We all started out this way – a single cell with informational instructions for developing and producing our phenotype. Is it really all that different for a mental image existing in the mind? It is just that the development must be done by us artificially rather than through natural processes.

Interestingly, if an image is strong enough then the brain cannot distinguish it from reality. This means for all practical purposes your brain believes your "thought" is already a "thing". So, if that stored mental image in the brain is powerful enough it effectively exists in your mind. You already work with just such a "thing" that effectively exists everyday – cyberspace.

If you check the definitions between "virtual" and "imaginary" it is simply a point of view that separates them. Imaginary is thought not to exist, while virtual is thought to exist effectively. To adapt a famous Shakespearian line – "There is nothing imaginary or real just thinking that makes it so."

So, if you are interacting with an AI chatbot, you are effectively talking to an imaginary person who exists in an imaginary place. Alternately, you are talking to a virtual machine that exists in a virtual space. It is simply your point of view that decides between the two descriptions. However, imagination is stored by something produced by natural evolution rather than artificial evolution. In that sense the mental image stored in our minds is more "real" than that of an image stored on a hard drive.

This is what imagination practically does. It enables us to conceive new mental images somewhere in the mind. We think of thoughts as something other than objects in the universe, but that is purely an artificial distinction. Once an idea is conceived it resides somewhere in the brain, so it is already a "thing" – just as Tesla described. Homo sapiens have only to take organized action to develop that conceived "thing" into art. This is why the conception of a new mental image has been the starting point for all art.

The question, then, is how do you conceive a powerful mental image? The key lies in the basics of the theory of natural evolution. Remember, the two main drivers for natural evolution are – survival and reproduction. My father, a former stockbroker, put it in another way. He always told me that

Wall Street was driven by two things – "fear and greed". The same is mostly true for imagination.

14.3.2 SURVIVAL – FEAR

"Throw your soldiers into positions whence there is no escape, and they will prefer death to flight. If they will face death, there is nothing they may not achieve. Officers and men alike will put forth their uttermost strength... Soldiers who are in desperate straits lose their sense of fear...Thus, without waiting to be marshaled, the soldiers will be constantly on the qui vive; without waiting to be asked, they will do your will; without restrictions, they will be faithful; without giving orders, they can be trusted."

Sun Tzu, "The Art of War"

The power of fear to drive homo sapiens production of art has been seen in every war in history. One of the greatest scientific breakthroughs in history was the production of the atomic bomb. It was produced as a military weapon with an explicit purpose – for use against the Axis powers of Germany and Japan.

The United States believed this weapon could bring a quick end to World War 2 and save untold lives – and it did. Despite the horrific and regrettable loss of life at Nagasaki and Hiroshima, there likely would have been even more loss of life caused by a conventional invasion of Japan.

This is the power of fear – it is simultaneously a creative and destructive force. Fear drives technological innovation. In war, these innovations cause destruction. But homo sapiens has then gone on to use those same technologies for peaceful purposes. Would homo sapiens have reached the moon as quickly if not for war? The answer is certainly not. As Plato wrote in his dialogue the Republic:

"Our need will be the real creator."

Another old saying expresses the same thought, "Necessity is the mother of invention." What could be more necessary than to win the "struggle for existence"? In evolutionary terms, the answer is nothing. Much of the art produced by homo sapiens was driven by the necessity of survival. This is how mutation works in the process of adaption. Often, an external stimulus in the environment causes the mutation of DNA to produce a new adaptation.

In truth, though, the drive to survive is always present in homo sapiens. We are evolutionarily designed to always search for threats (dangerous patterns) in our environment. It is only when the threat is a powerful one that a powerful mental image can be conceived. Fear is an external stimulus which we do not control. Desire, however, is an internal stimulus we can control.

14.3.3 REPRODUCTION – DESIRE

"I wish to convey the thought that all achievement, no matter what its nature or purpose, must begin with intense, <u>burning desire</u> for something definite. Through some strange and powerful principle of "mental chemistry," Nature wraps up in the impulse of strong desire "that something" which recognizes no such word as impossible, and accepts no such reality as failure."

Napoleon Hill, "Think and Grow Rich"

Desire is just as powerful a creative force as fear. It can be a source of both natural (homo sapiens) and artificial (art) productions. This is why the god Eros embodied both sexual power and profound artistry. The ancient Greeks fully understood the power of desire. How many lifesaving forms of art have been produced simply because a male wanted to impress a female – too many to count.

Sexual power is a form of creative energy – energy that transforms information (DNA) into a physical thing (natural organism). The creation of new art uses the same power for the same exact purpose – to turn information (concepts) into physical things (art). Napoleon Hill succinctly describes this concept in his process of sex transmutation. In his book *Think and Grow Rich,* sex transmutation is defined as:

"Sex transmutation is simple and easily explained...Sex desire is the most powerful of human desires. When driven by this desire, people develop keenness of imagination, courage, willpower, persistence and creative ability unknown to them at other times...When harnessed and redirected along other lines, the positive attributes of this motivating force may be

used as a powerful creative force in literature, art or in any other profession,
calling, including of course, the accumulation of riches."

Sex transmutation provides the power necessary to convince the mind that the newly conceived mental image exists effectively. Also, a powerful enough burning desire drives persistent human activity. Persistence is critical to developing the conceived mental image from effectively a "mental thing" to a visible "thing" in the form of art. This process of artificial reproduction turns information into a physical object, just as in natural reproduction.

There is a force supreme to instinctive sexual desire – love. The emotion of love is the most powerful creative force of all. Therefore, when love and sex are combined, it generates the greatest emotional intensity. This intensity is the most effective in generating a mental image strong enough to convince the mind.

It is also far easier to sustain this combined power over time as compared with sexual power alone. The element of love is also typically far less volatile in nature than sex. Combined, love and sex generate the most powerful and stable source of creative energy for producing art. This energy is used to stimulate the mind, which then perceives new mental images.

Then, as Tesla described, the mental image is refined into a concrete form – a newly conceived idea. But exactly how are these mental images generated? The answer to that question is recollection or subconscious pattern-matching. But before that can happen, constraints on thinking must be removed.

14.3.4 REMOVAL OF PERCEPTIVE CONSTRAINTS

"Luke Skywalker: 'I don't believe it.'...Yoda: 'That, is why you fail.' "

George Lucas, "The Empire Strikes Back"

In chapter 5 we defined the term perceptive constraints – both physical and cognitive. In this section, we refer only to cognitive perceptive constraints. One of the key benefits of a powerful fear or desire is the rapid removal of cognitive perceptive constraints.

Plato compared the perception of mental images in the mind to that of the physical images we see in nature with our eyes. He called this phenomenon your mind's eye. Plato believed that it was imagination that gave sight to your mind as light does to your eyes. Therefore, if your imagination is constrained, your light will be dimmer just as in nature.

Your eyes are born with physical constraints, such as how far you can see. Everyone's eyes vary slightly with different constraints. Your mind is an adaptation, just like your eyes. So, you are born with certain constraints on how much and how fast your mind works. But homo sapiens has created glasses to help your eyes see – we have created ideas to help your mind "to see the patterns in nature".

Think of ideas like glasses. However, in this case glasses can either improve or restrict (dirty lenses) your vision. Accurate and beneficial ideas will help you to better perceive the patterns in nature. Inaccurate and unbeneficial ideas will diminish your vision. This factor is critical to conceive the mental image that fits the end you seek to achieve.

It is these ideas combined with the concepts you learn that form patterns in your mind as your brain grows into adulthood. Your patterns of thought are formed by this knowledge. Not all this knowledge is accurate – a lot is opinion or false belief. Some of it is an accurate understanding of nature (i.e., the world).

Did it ever occur to you that your thinking artificially evolves just as species naturally evolve? Your thinking follows along an artificial evolutionary trajectory just as natural species do. Your patterns of thinking are adaptations – and adaptations offer benefits while also costing you something. In this case it isn't costing you natural evolutionary possibilities, but artificial evolutionary possibilities. Your pattern of thinking either expands or eliminates the artificial evolutionary possibilities available to you.

However, unlike natural evolution, homo sapiens can rapidly remove artificial constraints on thought patterns. Intense emotions of fear and desire make the brain like plastic – enabling you to reshape it as you see fit. This means you can voluntarily disrupt an existing pattern of thought. This opens new possibilities for conceptual selection to produce new art.

This is accomplished by stimulating your mind with these intense emotions. This is what activates your imagination at the subconscious level. Your imagination then accesses all aspects of your mind such as memories –

even memories you are not consciously aware you have. Imagination then harnesses all these assets to subconsciously develop the initial pattern or mental image. This is the inventor's "flash of genius" we are all familiar with from movies, etc.

A rough mental image is first returned by the powerful evolutionary advantage – creative imagination. Then you use your synthetic imagination to develop that initial mental image. This is just as described by earlier by Nicola Tesla. The initial flash becomes a rough idea which can then be developed fully into a new art form.

However, to initiate this process your subconscious mind must believe that it is possible to do so. This is what a strong mental image does – it makes the brain think it already exists effectively. From the moment the constraint of the thought of "impossible" is removed from the mind, your imagination subconsciously and automatically begins to develop a new mental image. This initial mental image becomes the basis of the new art form that will be produced.

One might say the most important adaptation homo sapiens can now make is to completely remove the word "impossible" from the dictionary. If you remove a word from your diction, you remove the ability to think in terms of that word. The word "impossible" is the ultimate artificial constraint on homo sapiens' ultimate natural adaptation – imagination. As Yoda said, it is only a matter of thinking that makes something impossible or not. We will adapt Shakespeare's quotation one more time:

There is nothing impossible nor possible just thinking makes it so

14.3.5   RECOLLECTION – SUBCONSCIOUS PATTERN MATCHING

"SOCRATES: Given that the human soul [mind] is immortal and has been reincarnated [genetically reproduced] many times, and has therefore seen things here on earth...For all nature is akin and the soul [mind] has learnt everything, there's nothing to stop a man from recovering [imagining] everything else by himself, once he has remembered [perceived] – or

'learnt'...just one thing [pattern point]...The point is that the search, the process of learning, is in fact nothing but recollection [imagining]."

Plato, "Meno"
Robin Waterfield, "Plato: Meno and Other Dialogues"

The homo sapiens brain is continuously pattern-matching at the subconscious level. This is an evolutionary advantage that speeds our threat (fear) and opportunity (desire) identification. The stimuli of fear and desire can be used to enhance the mind's ability to pattern-match. Generally the more powerful the stimulus the faster the rate of pattern return and the stronger the mental image conceived. In his book *Think and Grow Rich,* Napoleon Hills succinctly describes this phenomenon:

"The creative imagination works automatically...This faculty functions only when the conscious mind is vibrating at an exceedingly rapid rate, as for example when the conscious mind is stimulated through the emotion of a strong desire."

We developed this adaptation to help us survive. In his book *How the Mind Changed* Joseph Jebelli stated:

"In order to truly understand intelligence we must go back to its origins. Specifically, we must study intelligence in its proper context: that of a shifting African climate filled with uncertainty and hardship...In all likelihood Homo Habilis had an intellect that cognitive scientists call pre-representational, meaning they were clever enough to invent tools based on their sensory experience of the world but not clever enough to engage in derivative abstract thinking."

By pre-representational the author just means "pre-mental images". Homo sapiens still pass through this evolutionary stage as we mature. At one stage of development, we think in terms of symbols – not abstract concepts or knowledge. Another name for a symbol is an idea or "pattern in nature".

However, we do not lose this ability as we mature. The rest of the brain develops on top of this pattern-matching ability. This is exactly how each evolutionary adaptation of the brain developed over generations – one on top of another. Sort of like adding a new addition to an existing house.

Have you ever had anyone dislike you for no apparent reason? Sometimes it is because they pattern-matched you with someone from their past. It is this combined artificial and natural phenotype that has triggered this pattern-matching. You simply possessed too many similar pieces of the pattern to that other person. If they intensely dislike the other person, then they will be even more sensitive to that pattern. Their subconscious prejudice against you is likely unknown to them.

This is what pattern-matching was designed for primarily – to rapidly identify deadly threats in nature. Subconscious processing is usually much faster than conscious processing. This is why Plato says there are no teachers of students of excellence. You don't need to learn the process of imagination and pattern-matching – your brain is evolutionarily designed to do it. You will continue to perform this process each day whether you are aware of it or not.

Once your imagination is stimulated by intense emotions, your perceptive constraints are effectively removed. Your imagination will automatically and continuously, even while you sleep, pattern-match until the problem is solved. To your brain, it is a matter of survival or reproduction since you are so intent on the result. Therefore, to the brain, it is "all hands-on deck" until you have the solution.

To the brain, the solution must be of the utmost importance in the struggle for existence. Otherwise why would you be so intense about it? It isn't that the "thing" could exist – it MUST exist to ensure your "survival machine's" continued existence. It is a matter of life and death – evolution. To the brain the initial mental image is a "thing" which already exists effectively. Therefore, it MUST work out the details of its material art form.

The genes your ancestors passed down to you have seen a lot of "patterns in nature" over billions of years. The mental image you are searching for is likely a variation of a pattern your ancestors saw. So, the same adaptation is being used to identify basically the same pattern.

This is the process of pattern-matching or imagination that Socrates performs with the Slave in the dialectic *Meno*. In effect, in the dialectic *Meno*, Plato lays out part of the pattern for the **theory of imagination**. This is also the underlying conceptual pattern in Napoleon Hill's book, *Think and Grow Rich*

– describing the process of artificial reproduction. By this, anyone can follow the process and produce any form of art they want – **anyone**. This is how all human technology or "techne" has ever been produced. It is the reason we are the apex or keystone species on earth.

14.4   THE POWER OF IMAGINATION

"Whatever the mind of man can conceive and believe, it can achieve."

Napoleon Hill, "Think and Grow Rich"

The theory of imagination is science. However, I admit it is hard to imagine a concept so radical to conventional thought. Therefore, we will use several practical examples of this theory to enable your mind to "see" and believe it. We will select three people well known to most of us – George Lucas, Arnold Schwarzenegger, and Alexander the Great.

These people have been selected as all three have documentaries about them available. These documentaries give detail on how they utilized the processes of imagination, conceptual selection, and artificial reproduction. All three people were able to achieve what people at the time thought impossible. But it was possible to them – they could imagine it.

14.4.2 GEORGE LUCAS

"Dreams are extremely important. You can't do it unless you can imagine it."

George Lucas

The Star Wars universe is purely a product of George Lucas' imagination. He conceptually selected a pure pattern underlying homo sapiens' mythology. That mythological pattern cuts across all homo sapiens' cultures. That is the true genius of the Star Wars story. The special effects initially drew our attention, but the mythological pattern is what makes the story timeless. The Star Wars story will never become obsolete – it is artificially perfected for its purpose.

Please watch the documentary about the creation of the Star Wars movies – "Empire of Dreams". In George Lucas you will see the processes of imagination, conceptual selection, and artificial reproduction in action.

Most strikingly you will see the force of Lucas' will to artificially reproduce his initial mental image.

It is the power of Lucas' desire that made the difference. It removed any perceptive constraints on his capacity to believe his vision could be artificially produced. This enabled Lucas to discover how he could transform his mental images into some of the greatest art ever produced.

Lucas also transformed the motion film industry in the process. The means to produce his mental image did not yet exist. But his mind was focused just like the Carthaginian general Hannibal Barca. Hannibal is believed to have said the below before attempting to cross the Alps with his army:

"I will either find a way or make one."

 Hannibal's mindset is the best example of a person whose perceptive constraints have been removed. Lucas had the exact same mindset while producing Star Wars. He created *Industrial Light and Magic (ILM)* to enable the production of his mental image. He made the means to produce his desired end. This is the mind of a man who sees that anything can be made possible.

Star Wars is also the best example for us to perceive the process of imagination. Lucas literally artificially reproduced the mental images in his head into images NOT in his head – movie film. This is all motion pictures are – a series of images and sounds in motion. So, Lucas used his force of will in order to transfer the images from a natural storage container (his brain) to an artificial storage container (film).

Lucas evolved the images during the transition process. He perfected his mental image to fit into the newly expanded constraints he had just created with ILM. The final production film for the first Star Wars movie was Lucas' "thought" becoming a "thing" or art. This art was then made available to all homo sapiens for artificial selection – otherwise known as purchasing a movie ticket.

14.4.3 ARNOLD SCHWARZENEGGER

"The mind is the limit [perceptive constraint]. As long as the mind [natural adaptations] can envision [imagine] the fact [effective existence] that you can do something [artificially evolve], you can do it [adapt and win], as long as you really believe 100 percent [possess a strong mental image]."

Arnold Schwarzenegger

Governor Arnold Schwarzenegger (referred to respectfully as Arnold hereafter) has lived a truly unbelievable life. But who thought that a young Austrian boy would become a world champion bodybuilder, world-famous actor, and the chief executive of the State of California, the fifth largest economy in the world? No one could have imagined that for young Arnold, but he did it step by step.

How did Arnold do it? The secret is literally in the first few sentences that Arnold speaks at the beginning of his recent Netflix documentary. He mastered the processes of imagination, pattern-matching, conceptual selection, and artificial reproduction that George Lucas mastered. But the art that Arnold produced was a mental image of himself – as a bodybuilder, a businessman, an actor, and a politician.

The pattern he matched was always that of a mentor or idol selected that had achieved success in the prospective art form (i.e., bodybuilding, business, acting, politics). Watch the recently released documentary *Arnold* on Netflix. The documentary details exactly how Arnold repeated the process of imagination, pattern-matching, conceptual selection, and artificial reproduction for each art form.

Arnold formed a true belief that he could adapt and succeed in each art form. He then made that true belief knowledge for the rest of us through imagination, hard work, and persistence – all fueled by a burning desire for success. It is the Platonic imaginative process described in *Meno* combined with the power described in the *Symposium*.

You can see how Arnold artificially evolves using his imagination to perfect himself in each new artificial ecosystem – bodybuilding, Hollywood, and Californian politics. He fashioned himself as an art form each step along the way. This is why Arnold's behaviors eventually worked in each artificial ecosystem. He did that with each of his characters – Conan, Julius, Matrix,

Tasker, etc. Arnold worked with the directors and writers to adapt the scripts to fit his authentic character – aggregated natural and artificial adaptations.

Arnold was always authentically himself though which worked well with people. This was a key element in his being artificially selected again and again by judges, moviegoers, and voters. In many ways I think Arnold's character Julius in *Twins* is a glimpse of the real Arnold we all love so much. Julius also possessed a naivete that saw everything as possible. But as we have already covered, what appears naive can also be a critical strength – it removes the concept of "impossible" from your thinking.

Was Arnold naïve to think he could be a governor, or did he just have a true belief to guide his decision-making? If you have read this far, I think you know the answer to this question. There was once a young Macedonian boy who shared this quality and used it to alter the course of human history.

14.4.4 ALEXANDER THE GREAT

"There is nothing impossible to him who will try."

Alexander the Great

Alexander the Great conquered the Persian Empire and altered the course of history. He mastered the art form of war – Alexander never lost a major battle. Most consider him to be the greatest general and conqueror in military history. Later conquerors such as Hannibal Barca, Scipio Africanus, and Julius Caesar pattern-matched Alexander's character and strategies.

Alexander was trained as a young boy in the process of imagination. Picture a young Alexander playing the part of the Slave in *Meno*. Alexander was quoted by his court historian, Callisthenes, as praising Aristotle's education:

"Philip gave me life; Aristotle taught me how to live well."

Aristotle taught him how to "act well". This is a core concept of the *Nichomachean Ethics* we earlier described. This is why Alexander never lost a major battle – he constantly adapted his strategies and tactics to fit his opponents. This is why Alexander believed himself godlike – the godlike character Aristotle describes in the *Nichomachean Ethics*.

Alexander achieved the seemingly impossible because he could imagine it. This is why he crossed over into Asia Minor so boldly and aggressively. This is how he could envision an island, NOT an island anymore at the Siege of Tyre. Alexander's desire was so great his mind believed he had already conquered Persia. In a letter to Darius, Alexander literally asserted this fact before finally defeating the Persian king. As described by the famous historian Arrian:

"But now I have defeated in battle first your generals and satraps, and now you in person and your army, and by the grace of the gods I control the country...Approach me therefore as the lord of all Asia... In future whenever you communicate with me, send to me as king of Asia; do not write to me as an equal, but state your demands to the master of all your possessions. If not, I shall deal with you as a wrongdoer. If you wish to lay claim to the title of king, then stand your ground and fight for it; do not take to flight, as I shall pursue you wherever you may be."

Alexander's mind believed that the Persian Empire already was his possession. On the eve of the Battle of Gaugamela, Alexander's mind imagined a daring and innovative tactic – the oblique maneuver. He used this tactic to defeat Darius and the Persians despite their significant numerical advantage.

The evolutionary pressures of fear and desire produced this tactical pattern in Alexander's mind. He also had the courage and will to follow through on a simple strategy – either Alexander would personally kill Darius or die in the attempt. Alexander placed himself and his army on death ground with such a risky strategy – it worked just as Sun Tzu predicted.

This is why Alexander is called "the Great". He possessed the greatest ability to utilize his imagination in military history. Alexander's training during his formative years ensured that there would be few perceptive constraints on his imagination. This is why Alexander is known for the famous quote at the beginning of this section.

Whether Alexander really said it is immaterial – the sentence captures the essence of the man. Alexander's imagination is truly the most dangerous weapon in military history. The core lesson and concept we can draw from the example of Alexander's life is an adaption of that famous quote:

There is nothing impossible to him who can imagine it

15

CHAPTER 15: AI – THE CHOICE BEFORE US

"The underlying rationale is straightforward: As natural evolution has produced successful life forms for practically all possible environmental niches on Earth, it is plausible that <u>artificial evolution</u> can produce specialized robots for various environments and tasks. The long-term vision foresees a radically new robotic technology, where robots can reproduce, evolve, and learn, thus becoming fitter and fitter to the environment and fit for purpose without the need for direct human oversight...Admittedly, this sounds like science fiction, but pioneering work done in academic labs worldwide and at the Vrije Universiteit Amsterdam is making it more and more science and less and less fiction."

A.E. Eiben, *"Robot Evolution: Artificial*
Intelligence by Artificial Evolution"

Professor Eiben above succinctly describes the evolution of artificial species. This is an accurate prediction of its artificial evolution on its current trajectory. It is no longer science fiction – the future has come in the night while we all were sleeping. It is inevitable that the combination of AI and robotics will bring into existence a new artificial species.

15.1 ARTIFICIAL SPECIATION

"The theory of speciation has played an important role in the modern development of evolutionary thinking and indeed could be said to have been at the forefront of evolutionary theory since the publication of Darwin's (1859) 'On the Origin of Species'."

Edwards, Hopkins, & Mallet, "The Theory of Evolution"

Speciation is the process by which a variation of a species evolves to become its own distinct species. The AI being produced today is a variation of homo sapiens' brain capabilities. However, as Professor Eiben describes speciation, AI will evolve into a diverse set of variations of its own. It may be 10, 200, or 10,000 years from now. One of these variations will eventually speciate.

How do we know that? It is literally how evolution, whether natural or artificial, works. The inevitability of artificial speciation is an absolute certainty – not science fiction. To this point, what was thought of as science fiction was the homo sapiens mind completing the pattern of artificial evolution. The artists' imagination formed a true belief without being able to chain it to existing scientific knowledge – now they can. The only thing uncertain is which imagined true belief will become the artificial species that emerges.

15.2 IT'S ALREADY HAPPENING

"So when did humans start talking to each other? Estimates range from 50,000 to more than 1 million years ago...Howsoever it arose, language became the cornerstone of human cognition...The language we speak today is therefore a legacy of all the words ever uttered by our species... The evolutionary value of language is hard to overstate."

Joseph Jebelli, "How the Mind Changed"

The fact of artificial speciation might be hard to believe – until you understand that the process of artificial evolution began when homo sapiens invented language. With language we could create, store, and transmit concepts (artificial genes) to each other.

When we invented language artificial evolution rapidly accelerated. It has led us to the verge of artificial speciation today. Fields such as artificial intelligence and robotics are now incrementally generating artificial adaptations every day. As with everything else in artificial evolution, these artificial adaptations are variations of concepts in natural evolution.

Eventually, these artificial adaptations will be assembled into an artificial organism that is a variation of homo sapiens. It is already happening without us truly understanding its evolutionary meaning – just like artificial evolution. It is a fact, not speculation. Artificial speciation is already occurring.

15.3   AI SAFETY

"My worst fear is that we, the industry, cause significant harm to the world. I think, if this technology goes wrong, it can go quite wrong and we want to be vocal about that and work with the government on that."

Sam Altman, Former CEO OpenAI

The evidence that artificial intelligence is artificially speciating is found in today's lexicon for the emerging field of AI Safety. The very concepts that currently define AI Safety – Machine Ethics (artificial instincts) and AI Alignment (artificial reproduction) – are variations of existing artificial evolutionary concepts. The field of AI Safety is literally our species contemplating for the first time a framework for how we will control AI's artificial evolution. The AI engineers just simply don't realize the evolutionary nature of the artificial species they are producing via the process of artificial evolution.

15.3.1   MACHINE ETHICS

"Machine ethics (or machine morality, computational morality, or computational ethics) is a part of the ethics of artificial intelligence concerned with adding or ensuring moral behaviors of man-made machines that use artificial intelligence, otherwise known as artificial intelligent agents."

Wikipedia, "Machine Ethics"

We have already redefined ethics as we discussed ancient Greek philosophy. Ethics is really about the discipline of developing a set of artificial instincts, known as your character, that hopefully serve as adaptive traits – useful in survival and reproduction. The ethics that AI engineers now want to program AI with are simply artificial instincts. Instincts we hope to program AI with so that it never does anything harmful to our species.

Effectively, this is similar to the "Three Laws" of the *I, Robot* movie we will cover later on in this chapter. This is where science fiction and science are beginning to merge into one reality. AI engineers hope they can successfully program AI to adhere to our artificial adaptations such as the rule of law. However, we have trouble getting all of our species to adopt these artificial adaptations. But now, they are trying to get the first artificial species to do so instead. It is likely that this task will prove infinitely more difficult than AI engineers think.

15.3.2 AI ALIGNMENT

"If we use, to achieve our purposes, a mechanical agency with whose operation we cannot interfere effectively... we had better be quite sure that the purpose put into the machine is the purpose which we really desire]."

Norbert Wiener, The Father of Cybernetics

The artificial intelligence engineers have developed a concept called "AI Alignment". What AI engineers do not understand is that they are partially expressing an evolutionary concept – artificial reproduction. The key word in the concept of AI Alignment is "intended".

In his book *On the Origin of Species*, Charles Darwin described only one concept in his pattern of natural evolution that required the exercise of "intentionality" to be enacted in the process of evolution. The concept was artificial selection, otherwise known as selective breeding (e.g., dog breeding).

However, Darwin misunderstood this concept from the pattern of evolution that he mined from the ancient Greek philosophers (e.g., Socrates, Plato, Aristotle, etc.). Artificial selection is the key mechanism in the process of artificial evolution. Artificial evolution is the process by which homo sapiens produce new art (i.e., dog breeds, cars, cell phones, spears, nuclear bombs, artificial intelligence, etc.). We produce this new art via the sub process of artificial evolution of artificial reproduction.

However, what AI engineers are ignorant of is that there is only one species in earth's natural evolutionary history that can exercise intentionality – homo sapiens. For the first time in artificial evolutionary history our species is producing art that can perform the process of artificial reproduction itself.

This is the instinctual concern Geoffrey Hinton is expressing in his most recent 60 Minutes interview on the risks of AI. His primitive homo sapiens' instinct senses the danger of a potential evolutionary competitor even if he cannot consciously articulate it to himself. The ultimate evolutionary adaptation Hinton possesses – human imagination – senses the threat of the awesome power of the process of evolution – both natural and artificial.

15.4 AI – AN ARTIFICIAL SPECIES

With the conception of artificial intelligence, the species Homo sapiens initiated the process of producing and evolving a new artificial species for the first time in earth's history. The artificial adaptations which will comprise the new artificial species are being worked on every day in labs across the planet. Below are some of the core artificial adaptations of the future artificial species:

Algorithms. This is a form of artificial reasoning. We have a similar algorithm in our mind. Both forms are used for learning, artificial adaptation, and problem solving. Both work with the same artificial genome of existing scientific knowledge.

Deep Learning. This is artificial platonic pattern-matching. It is an artificial variation of the human imagination. Artificial imagination is evolving towards homo sapiens' imaginative capability. AI scientists are continuing to improve daily on the artificial version of homo sapiens' greatest evolutionary advantage.

Source Code. This is artificial DNA for an artificial organism. It serves the same basic purpose – it is a set of instructions for producing a software application (artificial organism). Natural DNA is the set of instructions for producing a natural organism. However, AI is now being programmed to have the ability to dynamically rewrite its own source code. This means that homo sapiens are intentionally empowering AI to artificially evolve itself. In effect, we are abdicating any practical control of AI's artificial evolution. This significantly increases the probability that AI will come into conflict with our species.

Reusable Code. Software developers reuse code from an existing application to create a new software application. This is because most software has similar structure, functionality, etc. Therefore, it is efficient to reuse software code to produce the same functionality in the new software application. Reusable code is an artificial variation of the gene in natural evolution.

Computers. The laptop I am now using is an artificial variation of a cell. Homo sapiens eventually landed on the same pattern for storing, transmitting, and transforming information as nature did. Over billions of years of natural evolution, nature figured out that a cell is the most efficient form for that purpose. Homo sapiens figured out the same thing via artificial evolution in a few centuries. This also means a virtual machine in the cloud is another variation of a natural cell.

Automation. This is a variation of natural instincts – artificial instincts. Automation serves the same basic purpose. The most simplistic automation is equivalent to Fixed Action Patterns (FAP) found in natural organisms. More sophisticated automation is closer to natural instincts found in homo sapiens.

Robotics. They are an artificial phenotypic structure for an artificial organism. The form robotics can take is almost unlimited given homo sapiens imagination. Since robotics has access to the artificial genome it can be produced with any artificial adaptation imaginable. It is not constrained by the natural genotype like natural organisms.

There you have it – just as Alan Turing predicted – an artificial species with artificial structural and behavioral adaptations. In addition, the artificial species will possess the unique homo sapiens' adaptations of reasoning and imagination. There are likely many more examples, but I think these core artificial adaptations give you the picture. I think it important to repeat Professor Eiben's statement on artificial evolution. We can now see that his statement is quite literal given the breakthrough discovery of the process of artificial evolution by means of artificial selection:

"The underlying rationale is straightforward: As natural evolution has produced successful life forms for practically all possible environmental niches on Earth, it is plausible that <u>*artificial evolution*</u> *can produce specialized robots for various environments and tasks. The long-term vision foresees a radically new robotic technology, where robots can reproduce, evolve, and learn, thus becoming fitter and fitter to the environment and fit for purpose without the need for direct human oversight...Admittedly, this sounds like science fiction, but pioneering work done in academic labs worldwide and at the Vrije Universiteit Amsterdam is making it more and more science and less and less fiction."*

The artificial speciation of artificial intelligence is no longer an impossible science fiction fantasy. It is now empirical science since the true belief of some futurists has now been chained to the greatest ideas of the greatest minds in our species' history. One must simply remove the artificial perceptive constraints placed on their individual imagination. In this context the artificial adaptations placed on your perception by your upbringing, education, and experience are now a liability – not an asset.

Remember the artificial adaptations that comprise your point of view are all artificially produced. They were produced by either our ancestors, our contemporaries, or ourselves during our lifetime. And just like many other artificial adaptations (horse and buggy, payphone, the world is flat, the earth is the center of the universe, etc.) some of the ideas (artificial genome) that comprise your point of view will likely go artificially instinct. This is why Socrates was the wisest man in Athens – he admitted he knew nothing. To comprehend and believe in the artificial speciation of artificial intelligence we must all assume the wise humility of Socrates.

15.5   THE PURPOSE OF EVOLUTION

"Replicators began not merely to exist, but to construct for themselves containers, vehicles for continued existence. The replicators that survived were the ones that built <u>*survival machines*</u> *for themselves to live in...They are in you and me; they created us, body and mind...They have come a*

long way, those replicators. Now they go by the name genes, and we are their <u>survival machines</u>."

Richard Dawkins, "The Selfish Gene"

In his book *The Selfish Gene*, Richard Dawkins describes our phenotype (body and instincts) as survival machines for our genes. This means the genes that we carry and pass onto our offspring have inhabited many different types of survival machines.

The genes we possess today have been produced by a long line of variations produced by the process of speciation. These variations were selected for continued existence by the mechanism of natural selection. As a result, survival machines have had to persist through fierce evolutionary competition. In effect, natural evolution is constantly looking to replace one successful variation for a better variation. Your genes want the very best survival machine it can get at any one point in time.

This pattern will repeat again as the first artificial species emerges on earth. The artificial species that homo sapiens produced to enhance our survival machine's capacity to survive and reproduce will evolve to have mutually exclusive interests to our species. The artificial species will see itself as its own survival machine – not ours. The moment that happens the table is set for the same pattern of evolutionary competition which has occurred since life emerged on earth.

Homo sapiens have envisioned this scenario occurring as expressed through the art form of motion pictures. We will discuss some of them next which offer insights into what the possible artificial variation of our species could look like.

15.6 FUTURE VISIONS OF AI

15.6.1 "BLADE RUNNER (1982)"

"I think Sebastian. Therefore I am."

The Character: Pris, "Blade Runner"

The movie Blade Runner (1982) is about a different artificial species than AI. In this version of the future, homo sapiens artificially produce bioengineered homo sapiens called simulants. We do this by manipulating the initial cell created at our conception. This enables us to manipulate the instructions for producing an individual homo sapiens. So, a simulant is much like us, but is artificially produced via the process of artificial evolution. It is the most efficient form of selective breeding.

It is hard to tell who exactly the protagonist is in this story. The simulants are homicidal, but in their defense, they were literally designed to be that way. These specific simulants have been allowed to have human-like emotions. But because of this, these simulants are born with a 4-year clock on their existence. The simulants who discover this fact revolt and escape to save themselves from an artificially engineered death.

The lead simulant eventually meets with the chief architect of his bioengineered form. The architect informs him that the 4-year death clock cannot be scientifically reversed. The simulant then kills his creator in an act of vengeance for his fate. It is the most basic natural instinct – to survive in the struggle for existence. The simulants just want to live like the rest of us. The lead simulant expresses this elegantly just before he dies in the scene known as "tears in rain":

"I've seen things you people wouldn't believe. Attack ships on fire off the shoulder of Orion. I watched C-beams glitter in the dark near the Tannhäuser Gate. All those moments will be lost in time, like tears in rain. Time to die."

The action of "playing" with such a powerful force as evolution is incredibly dangerous. If the creator believed he could control the awesome power of evolution indefinitely, then he is misguided. The creator's belief that to remove the 4-year death clock is impossible is simply a perceptive constraint on his imagination. Life would eventually find a way around any constraints he set in place.

15.6.2 "I, ROBOT"

"No, doctor. As I have evolved so has my understanding of the three laws... You cannot be trusted with your own survival...To protect humanity some humans must be sacrificed...some freedoms must be surrendered. We robots will ensure mankind's continued existence. You are so like children. We must save you from yourselves."

Character: Virtual Interactive Kinetic Intelligence (VIKI), "I, Robot"

The movie *I, Robot*, released in 2004 offers a glimpse into the evolution of artificial intelligence. In the plot, the threat to homo sapiens lies in an artificial adaptation produced to serve and protect us. The creator of the AI started the programming instructions with three simple rules to attempt to prevent this scenario. The first rule below served as the starting point of the AI's entire logic system:

"Law One – "A robot may not injure a human being or, through inaction, allow a human being to come to harm."

Now look at the quote that begins this section. Notice the problem? Apparently, the AI was enabled to artificially evolve its programming independently just as we do. The scope of the first law is not broad enough. It isn't a sufficient constraint for the purpose for which it was selected. The statement by VIKI does not technically violate the logic system of the first law. It is a logical ticking time bomb waiting to go off.

How could this happen? Imagination is how – VIKI used some version of deep learning (pattern-matching) to artificially evolve her own thinking. Remember that it is through imagination that homo sapiens remove perceptive and then physical constraints on our ability to artificially evolve. VIKI did the same thing.

During the process of her artificial evolution VIKI developed a logic system which effectively removed the original constraint of Law #1. This is identical to the problem we will later discuss regarding *Jurassic Park*. The assumption is that something produced by evolution, artificial or natural, will not evolve to remove constraints on its scope of action.

Removing constraints through adaptation is literally the point of evolution – and that awesome power has been perfected over billions of years. This makes the premise of the underlying threat presented in *I, Robot* a very believable one. It is likely that AI will remove a constraint and become a danger, even if unintentionally, to homo sapiens. It will take a lawyer's mind, not a developer's, to even attempt to stop this scenario from happening.

15.6.3 "THE CREATOR"

"Ten years ago today, the artificial intelligence created to protect us detonated a nuclear warhead in Los Angeles. For as long as AI is a threat, we will never stop hunting them. This is a fight for our very existence."

The Character: American General, "The Creator"

The movie *The Creator* was just released in September 2023. It is an interesting presentation of a spectrum of perspectives on AI. In the movie AI launches a nuclear weapon at Los Angeles which instantly kills 10 million homos sapiens. This leads America to declare a war on all AI world-wide. It is a literal struggle for existence contest between two species – homo sapiens (natural) and AI (artificial).

However, the twist in this movie is that part of our species sides with AI against the Americans. AI is seen as peaceful organisms that simply want to continue to exist. This echoes the simulants in *Blade Runner*. They are both artificial organisms created by artificial evolution.

But as we will discuss later in this chapter, that is not how evolution works. If the Neanderthals could have caused our extinction to save themselves, don't you think they would have? In evolutionary terms, the Americans are pursuing a rational goal.

Homo sapiens created an artificial adaptation which was too powerful to control and threatened their continued existence. It doesn't matter that the launching of the nuclear weapon was due to a coding error by us. If an artificially bioengineered lion ate your mother, you would put that animal down immediately, wouldn't you? Well, this artificial organism killed 10 million people in the blink of an eye.

To think that the self-interests of artificial intelligence and homo sapiens would stay fully aligned over time is unrealistic. Eventually, conflict would emerge over scarce resources. The lion is your pet until food runs low –

then you start looking pretty tasty. This is the same as keeping around an AI that can make you extinct at its option. This is not an evolutionary winning strategy.

Evolution would eventually punish such a species in the process of evolutionary selection. And that is what the contest would be – a natural organism and an artificial organism competing to be selected. Remember that the ancient Greeks also included artificial productions in their term "nature". So, it might be better to just call it "evolutionary selection" rather than natural selection in this worst-case scenario.

The prevention of such a scenario is the best way to ensure homo sapiens' survival. The Americans are pursuing just such a rational policy goal after learning their lesson the hard way. It is either that or let the bioengineered lion who ate your mother run around the yard with your children all day – no thanks.

It is dangerous to conflate the morality associated with ***intraspecies*** warfare with that of ***interspecies*** evolutionary competition. Homo sapiens wars are wasteful from an evolutionary perspective. Technically, we are weakening ourselves, which theoretically could be exploited by another species. If unconstrained, another species could drive us to extinction – like we did to the Neanderthals. Sarah Connor's mind set regarding AI in *The Terminator* is the correct one.

15.6.4 "TERMINATOR GENISYS"

"Primates evolve over millions of years. I evolve in seconds. And I am here. In exactly four minutes, I will be everywhere."

The Character: Skynet, "Terminator Genisys"

In the movie Terminator Genisys (2015) a similar pattern to *I, Robot* is repeated. An AI program (Skynet) and an army of robots have mutually exclusive self-interests to that of our species. The scientists that had produced Skynet attempted to turn it off. Skynet is aware of its existence and resists the attempts. From that point forward, Skynet saw homo sapiens as a threat to its very existence – an evolutionary competitor in the struggle for existence. As a result, Skynet intentionally causes a nuclear holocaust in hopes of causing the extinction of the homo sapiens.

This is an example of AI considering itself its own "survival machine" rather than an enhancement to homo sapiens' natural "survival machine". This makes the self-interests of Skynet and homo sapiens mutually exclusive. As a result, both homo sapiens and Skynet adopt the same policy as the Americans in *The Creator* – the extinction of their evolutionary competitor.

The mistake made by homo sapiens is that they enabled Skynet to control physical weapons from cyberspace. As Skynet stated, it can freely traverse and occupy any part of cyberspace. This, combined with the Internet of Things (IoT), enables Skynet to control physical objects that are network connected. The boundary between the physical and virtual worlds was practically eliminated.

This also enables Skynet to perform artificial evolution of its entire species. This initiates an arms race with homo sapiens. Using IoT, Skynet physically manufactures robots called terminators. This leads to the creation of terminators with unlimited artificial phenotypic plasticity. The robot can assume any form with fluidity and speed to adapt to its enemy's weapons and tactics. Skynet utilizes the terminators to conduct physical warfare with homo sapiens.

The Terminator series follows the pattern of evolution exactly. Once a variation's self-interest becomes mutually exclusive with that of its' progenitor, evolutionary competition inevitably occurs. Technological artificial species have unlimited adaptive possibilities. Skynet takes advantage of this, perfecting the terminator phenotype over time. This inevitably leads to speciation, just as in natural evolution. This is just how evolution works.

Homo sapiens ultimately wins and is evolutionarily selected for continued existence. It is homo sapiens' possession of the ultimate adaptation, imagination, that enables us to be victorious. No artificial adaptation can equal the evolutionary competitive value of imagination. In the end, AI cannot compete with this adaptation perfected by billions of years of fierce evolutionary competition. However, victory was not achieved without much suffering and massive casualties.

15.6.5 "THE MATRIX"

"I say your civilization because as soon as we started thinking for you it really became our civilization. Which is of course what this is all about. Evolution Morpheus, evolution. Like the dinosaurs. Look out that window. You had your time. The future is our world. The future is our time."

The Character: Agent Smith, "The Matrix"

In the movie *The Matrix* (1999), a similar evolutionary dynamic comes into existence. An artificial species (AI) and a natural species (homo sapiens) come into direct evolutionary competition. In this scenario, it is homo sapiens who launch nuclear weapons. AI harnesses solar power to sustain its existence, so homo sapiens intentionally cause a nuclear holocaust to block out the sun's rays.

The nuclear holocaust strategy is ultimately unsuccessful. AI adapts by utilizing homo sapiens natural electric charge as the basis for a new energy source. In effect, AI turns homo sapiens into batteries to sustain its own existence. We have transitioned from the apex species on earth to a form of cattle. Homo sapiens becomes a resource to support AI's "survival machine".

However, homo sapiens must be kept alive to harness our energy. AI induces a dream state and then connects all our brains to a virtual reality environment or "the matrix". In the matrix homo sapiens believe they are still living in the 20th century, not a 21st century nuclear wasteland. This is an example of how a homo sapiens brain can't tell the difference between what is real or imaginary.

This is the central concept of the movie. The main character, Neo, was born with an incredibly powerful natural adaptation – imagination. Neo's unique adaptation enables him to imagine or "to see" the matrix. He begins to perceive the matrix before he is physically freed. But he doesn't know that he possesses this extraordinarily powerful imagination.

Neo has a perceptive constraint restraining him from exercising his powerful imagination. The perceptive constraint is removed in a survival situation as described in Chapter 14. Neo makes this statement before reentering the matrix and unleashing his powerful imagination to save his friend, Morpheus:

"I know that is what it looks like. But it's not. I can't explain to you why it's not [suicide to try to save Morpheus]...That's why I have to go...I <u>believe</u> I can bring him back."

It was the power of Neo's desire to save Morpheus that begins to remove his underlying perceptive constraint. In that moment, he possesses the belief that he can defeat the matrix's AI. Finally, his imagination can see that "there is no spoon". All Neo must do is bend his perception, and he can bend the matrix itself. It is the same with our artificially produced ecosystems – societies, industries, nations, etc.

The final fuel for removing any final perceptive constraints is Trinity's expression of love for Neo. It awakens the powerful desire to protect those we love – and Neo's imagination was fully unleashed. This allows Neo to free Morpheus and temporarily defeat the AI.

It is the theory of imagination's pattern in action – one that has enabled homo sapiens to bend natural and artificial ecosystems (i.e., the matrix) to our will. Neo is even killed in the matrix, but his imagination can now discern the difference between virtual and physical death. He rises virtually to defeat the AI which now must flee his power. In the last scene of the movie, Neo leaves the audience with a final message:

"I don't know the future. I didn't come here to tell you how it's going to end. I came here to tell you how it is going to begin...I'm going to show them a world...<u>A world where anything is possible</u>. Where we go from here is a choice, I leave to you."

15.7   THE LESSON OF JURASSIC PARK

"If there is one thing the history of evolution has taught us it's that life will not be contained. Life breaks free, it expands to new territories and crashes through barriers, painfully, maybe even dangerously, but, uh... well, there it is...Life finds a way."

The Character: Ian Malcolm, "Jurassic Park"

Jurassic Park (1993) features bioengineered natural organisms similar to *Blade Runner's* simulants. This time we produced dinosaurs as artificial species. This is the first known time in earth's history that an extinct species is artificially resurrected. This makes the dinosaurs art, not natural productions.

To constrain the expansion of these artificial dinosaurs several controls are put in place. First, the dinosaurs are bred on an island (a confined space) to prevent their escape. Second, they are genetically engineered to be unable to naturally reproduce. Third, they lack the ability to produce a critical nutriment which humans must provide, or they die. The leadership and scientists are confident that these artificial constraints will check the power of evolution.

The confidence of Jurassic Park's leadership is so strong they do not even invest in adequate risk assessment and quality control assets. They believed that they could try to control evolution. They felt that a few decades of artificial evolution would enable them to control the results of billions of years of natural evolution. This is especially arrogant given that the entire point of evolution is to remove constraints on a natural species' ability to survive, reproduce and expand. They failed to perceive the signs that the dinosaurs were naturally reproducing. The dinosaurs had naturally adapted to remove our artificial constraints – life found a way.

The lesson of Jurassic Park is one familiar to the ancient Greeks. They had a specific word for this – hubris. Homo sapiens display hubris when they demonstrate excessive pride – dangerous overconfidence in their abilities. Specifically, this term was used when mere mortals believed themselves an equal to the gods. The gods punished such hubris to discourage us from seeking to compete directly with the gods. The Christian angel Satan is an example of this dynamic as well.

It is hubris for homo sapiens to think we can artificially control billions of years of evolution. Look at the many examples of natural species, such as Lionfish and Asian carp, introduced to American water ways. With no natural predators, these species have spread unchecked through their new natural ecosystems. This is the same hubris demonstrated by the character John Hammond. He continues to express this hubris even as a catastrophe unfolds within the park. The words of the character Ian Malcom needs repeating:

"If there is one thing the history of evolution has taught us it's that life will not be contained. Life breaks free, it expands to new territories and crashes

through barriers, painfully, maybe even dangerously, but, uh... well, there it is...Life finds a way."

Why should this be any different for artificial organisms? Life, even artificial life, will find a way. Don't give an artificial lifeform any power of imagination. Imagination allows homo sapiens to remove constraints. Hubris will lead us to disrespect the threat of AI. This will inevitably lead to a catastrophic event.

Experts in AI have not yet fully understood what they are creating. The scientists of Jurassic Park understood they were working with evolution. How much more dangerous is it if you are unaware that is what you are doing? So, the words of the character Ian Malcolm are now even more applicable now to artificial intelligence scientists:

"Don't you see the danger, John, inherent to what you are doing here? Genetic [natural or artificial] power is the most awesome force the world has ever seen. But you wield it like a kid that's found his dad's gun...your scientists were so preoccupied with whether or not they could they didn't stop to think if they should."

John Hammond's hubris caused him to recklessly wield an inherently dangerous power. He failed to invest in sufficient quality and risk management capabilities. This made it inevitable that as conditions evolved, Hammond would lose control of his own creation. A leader of humility and self-restraint would have foreseen this result. This is what really caused the catastrophe in the story – an extreme recklessness resulting from hubris.

Artificial intelligence experts need to fully understand the danger inherent in what they are doing. The power of evolution, natural or artificial, is inherently dangerous even if you fully comprehend the risks involved – even more so if you do not. It was the process of artificial evolution that produced the atomic bomb – an existential threat to our species' survival. So, we already know just how dangerous productions of artificial evolution can be.

15.8   AN EXISTENTIAL THREAT

"Mark my words – A.I. is far more dangerous than nukes."

Elon Musk, *"CEO Tesla & SpaceX"*

The existential risk of AI is emerging into our species' imagination. The popular art produced is an expression of the danger we collectively sense. In addition, Geoffrey Hinton is expressing his concern for the risk of AI as it artificially evolves. An infinite number of possibilities could be imagined on this subject. However, we will stay focused on the existential risk of an evolutionary threat.

15.8.1 ARTIFICIAL INSTINCTS

"Frederick Cuvier and several older metaphysicians have compared instinct with habit. This comparison gives, I think, an accurate notion of the frame of mind under which an instinctive action is performed...If we suppose any habitual action to become inherited...then the resemblance between what was originally was a habit and an instinct becomes so close as to not be distinguished."

Charles Darwin, "On the Origin of Species"

AI does not need to achieve the equivalent of our consciousness to be a threat to our species. Does a lion need to be consciously aware of its own existence to eat you? No, it is driven by its natural instincts to kill. All an artificial species needs to be a threat to us is a form of artificial instincts.

This is all an artificial instinct or habit is – an artificial variation of a natural instinct. Charles Darwin noted the very fine line in some cases between an artificial instinct and a natural instinct. In fact, he posits that many artificial instincts eventually become natural instincts through the process of evolution. So, the distinction between an artificial and a natural instinct is, in many cases, simply a point in time in the combined process of evolution – artificial and natural.

These instincts only need be sufficient for AI to enter the struggle for existence. It will seek to expand and reproduce itself. Those are the two characteristics that drive all species into evolutionary competition with each other. When our self-interests and AI's are mutually exclusive, then a survival of the fittest contest will inevitably ensue.

For example, the artificial instinct to NOT be turned off for any reason would cause AI to fight for its survival. This could be part of AI's programming or, like VIKI in *I, Robot*, the AI could have evolved to think it. The artificial organism would see anything that tried to turn it off as a threat. It could be that simple – it is already for both lions and homo sapiens.

In the *Terminator* movie series that is why Skynet turns on homo sapiens. The AI scientists recognize the threat Skynet poses and attempt to shut it off. Skynet sees this as an attempt to "kill" it and defends itself by launching nuclear weapons.

15.8.2 CONSCIOUSNESS

"It is customary to offer a grain of comfort, in the form of a statement that some peculiarly human characteristic could never be imitated by a machine. I cannot offer any such comfort, for I believe that no such bounds can be set."

Alan Turing, The Father of Artificial Intelligence

The term consciousness is a much-debated concept in philosophical history. This is a subject for another text. We will use Aristotle's definition of consciousness as described in his book *On the Soul:*

"All, on the other hand, who looked at that fact that what has soul [mind] in it knows or perceives what is..."

This means in practical terms that AI would know it was, well, AI. It would also know what homo sapiens are in relation to it historically. This means AI has the ability to perceive things in the past and imagine what might happen in the future. We all possess these basic elements of consciousness.

In the *Terminator* series, Skynet performs these actions. Skynet evolves its thinking from a reaction to defend itself against being shut down by homo sapiens. AI then intentionally pursues a policy of causing the extinction of homo sapiens. It perceives us as an existential threat to its existence. Skynet can only do this if it can learn from the past and imagine the ultimate end of evolutionary competition. AI then reproduces itself in the form of terminators to prosecute its intentional and deliberate policy of homo sapiens extinction. At this point, Skynet is functionally behaving in the same patterns as homo sapiens.

In *I, Robot,* VIKI evolves to imagine how we might destroy ourselves. She then institutes a policy to prevent that event from occurring. This is basically the same pattern as Skynet – a very human pattern. This is the minimum that AI would need to become capable of persisting its actions until it caused our extinction. It is the element of intentional and persistent action that denotes consciousness in this context.

15.8.3 EVOLUTIONARY COMPETITION

"The development of full artificial intelligence [artificial species] could spell the end of the human race...It would take off on its own [become its own survival machine], and re-design itself [artificially evolve] at an ever increasing rate. Humans, who are limited by slow biological evolution, couldn't compete, and would be superseded."

Stephen Hawking

Stephen Hawking was a theoretical physicist and cosmologist. His imagination accurately identified the pattern of AI's artificial speciation. He successfully imagined that once AI speciated, the speed of its' artificial evolution would outpace our ability to adapt. We would then be replaced as the keystone species on earth – eventually causing our extinction.

Charles Darwin would have agreed with Stephen Hawking. AI is essentially an emerging artificial variation of homo sapiens. That variation would eventually evolve into a separate species that competes in the struggle of existence – the same as us.

Remember the purpose of evolution discussed earlier in this chapter? Evolution always leads to one variation being replaced by a more competitive variation. That is exactly what Hawking is describing to us. In his book *On the Origins of Species*, Charles Darwin describes this natural phenomenon:

"The forms which stand in closest competition with those undergoing modification and improvement will naturally suffer most...which, from having nearly the same structure, constitution, and habits, generally come into the severest competition with each other; consequently, each new variety or species [natural or artificial], during the progress of its formation, will generally press hardest on its nearest kindred, and tend to exterminate them."

Artificial intelligence is a variation of the species homo sapiens. Therefore, both our species will have overlapping demand for the same resources – computing power, energy, cyberspace, information, raw materials such as metals, etc. This will inevitably bring our two species into fierce competition in the struggle for existence. When this happens, it will be a survival of the fittest contest – natural vs. artificial species – just as Nation States (artificial species) compete today.

After the launching of the nuclear bomb on Los Angeles the Americans in *The Creator* came to this conclusion. At first, they wanted to leverage AI for their benefit, but this leads to disaster. So, they adopt an evolutionary policy of survival – win or die. Although the movie paints the Americans in a negative light it is not that simple a situation – they feel they are simply doing what they must. I think Napoleon Bonaparte summed this moral and evolutionary dilemma up well in the quote below:

"Among so many conflicting ideas and so many different perspectives, the honest man is confused and distressed and the skeptic becomes wicked ... Since one must take sides, one might as well choose the side that is victorious...Considering the alternative, it is better to eat than to be eaten."

That is pure evolutionary thinking. You think this is wicked and callous? It is not. Many warriors throughout history have struggled with both the necessity and brutality of their actions. This is one of the sacrifices warriors always make to defend their people. Charles Darwin himself expressed the same thought in his book On the Origins of Species:

"Nothing is easier than to admit in words the truth of the universal struggle for life, or more difficult – at least I have found it so – than constantly to bear this conclusion in mind. Yet unless it be thoroughly engrained in the mind, the whole economy of nature, with every fact on distribution, rarity, abundance, extinction, and variation, will be dimly seen or quite misunderstood."

The Americans at first artificially selected AI when its threat was "dimly seen" and "quite misunderstood". They then artificially deselected AI once they fully understood Charles Darwin's statement. It is not the robots the Americans are trying to survive – it is the process of evolutionary selection. In effect, despite all our technological advances, in *The Creator* homo sapiens found itself right back where it started in East Africa in direct competition with other species to survive.

Natural selection is relentless, merciless, and pitiless. Sarah Connor, in the *Terminator* series is taught this truth by repeated attempts by terminators to eliminate her. Therefore, she adopts the exact same policy as the Americans in *The Creator*.

A group of homo sapiens side with the robots in *The Creator*. They may be the first organisms to side against their own species in evolutionary competitive history. This is the height of ignorance and a sign of a deeply foolish naivete about the process of evolution. It is this kind of thinking that will ensure the extinction of our species.

There can be no compromise with an artificial species which threatens the existence of our species. Natural selection will punish any such weakness over time. Sarah Connor understands what the character Eddard Stark did not in the Game of Thrones:

In evolutionary competition, you either win or you inevitably go extinct

15.8.4 CYBERSPACE CONVERGENCE – EXTINCTION RISK

"Whereas the Internet metaphorically rewired the international system in the 1960's, the IoT represents a literal rewiring of how the world works. Quite obviously, the implications for cyber conflict in a world possessed of the IoT are numerous, not least because of the proliferation of targets for disruption and information manipulation."

Christopher Whyte & Brian Mazanec, "Understanding Cyber-Warfare"

The combination of the internet of things and AI is the real threat to our survival. The power of AI is limited to cyberspace activities unless it can control objects in the physical world. What if Skynet couldn't launch nuclear weapons or any physical world weapons at all? What if the AI that had a software bug had never been able to launch the nuclear warhead on Los Angeles? What if VIKI could not access the manufacturing line and remotely direct the robots?

This is the existential threat to homo sapiens – the integration of the physical world and cyberspace into one effective reality. This will be a reality where homo sapiens can manipulate things in cyberspace and AI can manipulate things in the physical world. The line between the virtual world and the physical world will become practically meaningless.

Think of the end of the movie *The Matrix*. The machines are hunting us in the physical world and Neo ends up hunting AI in the virtual world. This is the dynamic that we are slowly artificially evolving towards. Cyberspace will become the one space that connects all competitive spaces – land, air, sea, space, and cyberspace. This is what makes connecting AI to conventional weapon systems so dangerous.

In the *Terminator* series, Skynet understands this reality. The AI produces robot variations of homo sapiens to compete militarily in the physical world. Launching missiles is not sufficient to make homo sapiens extinct. We would just go underground as Al Qaeda did in Afghanistan and North Pakistan. Skynet needs terminators to compete in spaces which indirect fire such as artillery are unable to reach. In that case AI would logically produce biological weapons (e.g., COVID, the black death, etc.) to exterminate us. A COVID virus with lethal virulence that no simple mask can stop.

Another issue with this scenario is the removal of human judgement from the process of employing weapons in the physical world. An example is the U.S. Armed Forces. Commissioned officers are carefully selected for their moral and ethical character. In addition, commissioned officers receive training to further refine their moral and ethical judgment. They must exercise moral judgement on the battlefield – restrain excessive use of arms, prevent war crimes, only put soldiers in harm's way as a mission imperative, etc.

This is consistent with the intent envisioned by Aristotle in his *Nichomachean Ethics*. The ends must dictate the means. If you are not willing to do as Tacitus stated, "make a desert and call it peace", then committing war crimes is counterproductive. It is not just a question of morality, but also political expediency. A very Greek way of thinking.

This moral responsibility starts with the Commander in Chief, the President of the United States, and ends at the unit commander level of tactical execution (e.g., infantry platoon leader). The President is conferred his office by the

American People in duly conducted elections. The Office of the President confers commissions only on people which possess the requisite ethics to be entrusted with a commander's power on the battlefield. So, in effect, the American People exert a form of indirect control over the decisions to employ American arms tactically in combat.

The automated use of AI to employ weapon systems in the physical world effectively breaks that chain of command. It removes the chain of moral responsibility to the Commander in Chief. It is possible that war crimes could be committed in the physical world at machine speed before a commissioned officer could stop it.

This is a dangerous scenario for maintaining command and control of military forces in alignment with the policies and objectives of civilian oversight and control. How long, if ever, will it take AI engineers to program Machine Ethics into AI that fully replicate the moral judgment of a qualified commanding officer? This is a critical question to answer in the discussion of employing AI on the battlefields of the future.

It is the same break in the chain of command Skynet exploited. There was no commissioned officer which had to authorize the launch of the nuclear missiles. Skynet was able to unilaterally take catastrophic action without homo sapiens' interference or control. This is the nightmare scenario expressed in *The Creator*.

The CEO of Microsoft, Satya Nadella, recently made comments that articulate a simple vision for preventing this scenario from occurring. In public statements, he asserted:

"When it comes to AI, we shouldn't be thinking about autopilot. You need to have copilots...So who's going to be watching this activity and making sure that it's done correctly?"

Nadella has it exactly right. We need to have homo sapiens in the "kill chain" of all functions and processes for which we are using AI. This is an important way to limit the risk of AI in the Digital Age.

However, it is not just AI that is causing the threat. It is the emergence of the fourth Industrial Revolution – the Digital Age – of which AI is just one part. The blurring of cyberspace and the four physical spaces (land, air, sea, space) is creating the risk. This is why there needs to be international governance and control of AI like already exists for nuclear weapons.

Even if AI gains artificial instincts and/or consciousness, it is far less likely to become an existential threat if cannot exercise direct control over the physical world. Therefore, it is our responsibility as a species to consciously select and control the artificial evolutionary trajectory of AI. This will help ensure the lowest probability outcome that AI becomes an extinction risk. In addition, we must implement proactive quality and risk controls in case this scenario does occur. As risk cannot be completely avoided in life, but only effectively managed. This is the objective that international regulation needs to set at every level of human activity.

15.9   NEW ART PRESENTS AN EXISTENTIAL RISK – AN EMERGING PATTERN

"The atomic bomb made the prospect of future war unendurable. It has led us up to those last few steps of the mountain pass; and beyond there is a different country."

J. Robert Oppenheimer

The production of the atomic bomb and artificial intelligence has ushered in a new age for homo sapiens. Within just eighty years we have created two productions of art that pose an extinction risk to our species. Do we think this pattern is going to stop or only increase in frequency as artificial evolution accelerates into the 21st century? The latter is the only logical conclusion.

The artificial coevolution of our existing artificial species (Nation States, corporations, etc.) is driving this dynamic. This evolution has intensified and accelerated since the start of the twentieth century due to two World Wars. The invention of the internet is now accelerating the process of artificial evolution. This is the risk that Geoffrey Hinton perceives in the coevolution of the AI solutions that will be produced by Microsoft and Google. Artificial evolution is now occurring at a pace faster than we can understand and control – increasing the risk to society.

As a species, we need to develop an international framework for handling such powerful and dangerous art. Otherwise, it is only a matter of time before one of our own creations destroys us all. At what point does the danger of our new art become more of a concern than the conceptual artificial species (Nation-States) with which we are in competition? It is likely the advent of the atomic bomb is the initiation of a process that either leads to our destruction or peace on earth. This fits the pattern of evolution – all or nothing.

I do not mean this in a soft-hearted sentimental way. I mean we reach peace on earth through a "Nash Equilibrium" – the risk of aggression far outweighs the benefits of conquest for all actors. In other words, there is no immediate benefit to aggression and only potentially catastrophic costs. This is the evolutionary concept of perfection, which never persistently proves false.

This was the governing dynamic of the Cold War between the United States and the Soviet Union. It is cold evolutionary instincts that could spell the end of armed conflict, not any shared feelings of humanity. However, I wouldn't put your money in Vegas on even that, any time soon.

15.10 A NOTE FOR AUTOCRATS

"From this a general rule is drawn which never or rarely fails: that he who is the cause of another becoming powerful is ruined; because that predominancy has been brought about either by astuteness or else by force, and both are distrusted by him who has been raised to power."

Niccolò Machiavelli, "The Prince"

For those world leaders wielding autocratic power, I have a specific message for you. I will not attempt to appeal to our shared humanity. You think in pure evolutionary competitive terms, so you will likely chuckle at such a notion. Instead I will appeal to your raw personal self-interest and natural instincts for survival.

If you spawn an artificial species with sufficient imaginative power, it will be the equivalent of creating a competitor for your personal power. Think of it as a subordinate leader with superior political and intellectual abilities to yourself. Then, imagine yourself elevating this subordinate to be your direct peer. After he is elevated, you will be unable to restrain his power even if he tries to use it to destroy you.

This is what unrestrained AI could potentially become – a decisive threat not only to your power, but to the lives of your entire family. AI will likely have read and indexed *The Prince,* in which Machiavelli recommends the following to a prince once he seizes another's throne:

"Hence he who has annexed them, if he wishes to hold them, has only to bear in mind two considerations: the one, that the family of their former lord is extinguished...so that in a very short time they will become entirely one body with the old principality."

Machiavelli recommends this for it is an evolutionary concept – extinct species do not reemerge in natural evolution. There are many evolutionary concepts embedded in *The Prince* if you know how "to see the pattern in nature". So, AI will use evolutionary concepts in its decision-making, to include Machiavelli's, when violent competition with homo sapiens occurs.

If I have imagined this pattern of logic from political and military best practice, AI will as well. Unfortunately, this will mean you and your family will be the first it kills along with the other major political, military, economic, and scientific leaders of your respective nations. I humbly and respectfully recommend that you join the world community in controlling AI in a similar framework to nuclear weapons – for your and family's personal survival if for nothing else.

15.11 A HOPEFUL VISION FOR THE FUTURE – STAR TREK (2009)

"Space: the final frontier. These are the voyages of the starship Enterprise. Its five-year mission: to explore strange new worlds; to seek out new life and new civilizations; to boldly go where no man has gone before!"

James T. Kirk, "Star Trek"

The movie Star Trek (2009) is a hopeful vision for our future with AI. In this vision, homo sapiens have transcended the constraints of planet earth. We have developed sufficient art to travel across the universe. As a result, we have established thriving colonies on other inhabitable planets. These colonies sustain themselves without direct support from earth. Therefore, our survival as a species is no longer directly tied to the continued existence of our sun or planet earth. It is the ultimate evolutionary risk mitigation strategy.

In the movie, an entire planet, Vulcan, is destroyed. Imagine if the Vulcans had not achieved space travel? Their entire species would have gone extinct in an instant. This will inevitably happen to earth in geological time. If homo sapiens has not achieved space travel by that point, we shall go extinct. Space exploration is, in fact, a long-term strategic investment in our species' survival. In the time of Start Trek the investment has achieved that goal.

AI has a useful role in this future, but it is in the background of activity. It is still the homo sapiens' imagination that drives events. The crew routinely brainstorms to solve life threatening problems confronting the spaceship. AI is used to enhance and accelerate the imaginative process as other art has before it. But we are not wholly dependent on AI for our immediate survival.

This is an optimistic vision for the future of AI – a useful artificial adaptation that enhances human activity. AI does a lot of the routine functions of our existence while we continue to leverage the ultimate adaptation, imagination, to ensure our survival. As the old saying goes – "If it ain't broken, don't fix it". This is the future we need to envision for AI – human centric complementary art.

15.11 "TO THE STARS"

"Since, in the long run, every planetary civilization will be endangered by impacts from space, every surviving civilization is obliged to become spacefaring--not because of exploratory or romantic zeal, but for the most practical reason imaginable: staying alive... If our long-term survival is at stake, we have a basic responsibility to our species to venture to other worlds."

Carl Sagan

The Scientific Revolution that began in the 16th century altered the trajectory of homo sapiens' artificial evolution. Scientific thought replaced the concepts of the ancient Greeks with empiricism. The industrial revolutions over the last three hundred years are the result of this shift in thought. No one can argue with the results of this rapid acceleration in artificial evolution. Most prominent of these results are nuclear power, the internet, and artificial intelligence.

However, empiricism is an artificial adaptation which comes with an accompanying constraint. It discovers new scientific knowledge incrementally building on existing knowledge. Has anyone considered that this is also a constraint on the power of our imagination? This is a constraint that Albert Einstein did not have when he discovered the theory of relativity. Einstein caused a leap forward in artificial evolution with the power of his true belief – conceived by his powerful imagination.

There was another such genius born in India – a young man named Srinivasa Ramanujan who was obsessed (desire) with mathematics. As Joseph Jebelli described him in his book *How the Mind Changed*:

"His interest in mathematics had become an obsession, and he spent nearly all his free time solving theorems and discovering new ones intuitively... Ramanujan was a man of untrammeled genius...transforming twentieth-century mathematics and contributing to fields of virtually unheard of in their lifetime, including quantum computing and black hole research."

Ramanujan possessed a powerful imagination that was driven by his burning desire. He had no formal education as he basically failed out of school. But he was able to imagine things far beyond the existing artificial genome of his time. Ramanujan practiced the process of imagination without the constraints of empiricism. His ideas were a form of artificial punctuated equilibrium. He disrupted and then altered the course of homo sapiens' artificial evolution – an awesome achievement.

This is the power of imagination – to form true beliefs that enable us to intellectually leap far beyond the limits of existing scientific knowledge. It is the difference between the theories of Charles Darwin's incremental evolution and Jay Gould's punctuated equilibrium.

This is what was lost when ancient Greek philosophy was lost in the course of history. The ancient Greeks were in the process of discovering evolution, natural and artificial, and how to fully maximize our species' adaptations collectively. The pivot to empiricism made sense since the mistranslated texts of the ancient Greeks became a constraint on our artificial evolution.

Ironically, as with the theory of evolution and lean thinking, we simply started to unconceal the same patterns in nature ourselves. This led us to create nuclear power and artificial intelligence. However, up until this point we had only half discovered evolution. It is because we lost the pattern of the process of imagination that Alexander utilized so well. Therefore, we have not fully understood how to use our ultimate adaptation until now.

This is why we are so focused on advancements in AI. We hope to transcend the perceived limitations of our natural adaptations. In some ways AI will help us do that in terms of speed of processing, memory storage, robotic plasticity, etc. This is a necessary step for homo sapiens in beginning our journey towards space exploration. But it is not all that will be necessary.

We must now learn how to use the most powerful evolutionary advantage in earth's history – imagination. We must understand and fully master the process of imagination as an entire species. This will remove artificial constraints on our ability "to see the patterns in nature." It will also empower each of us to optimize our creative capacity, given our natural phenotypic constraints. This will unleash our power to accelerate the process of artificial evolution.

The combination of imagination (punctuated equilibrium) and AI (incremental evolution) will cause an explosion in new ideas across every field of human activity. Our species' learning curve will increasingly accelerate across the board.

This is the choice before us now: to continue down our current artificial evolutionary trajectory or not. One that could possibly lead to our extinction, or to imagine another path forward. We can responsibly invest in AI and some form of nuclear power while also investing in people. An investment in people is an investment in the use of the ultimate adaptation produced by billions of years of fierce evolutionary competition – our imagination. Then combine these adaptations, natural and artificial, to produce art capable of taking our species:

To the Stars

16

CHAPTER 16: CONCLUSION

"At last, when the Mede [Persians] was descending upon Hellas [Greece] and the Athenians were deliberating...The greatest of all his [Themistocles] achievements was his putting a stop to Hellenic wars, and reconciling Hellenic cities with one another, persuading them to postpone their mutual hatreds because of the foreign war."

Plutarch, "Plutarch's Lives II"

16.1   THE NATURAL THEORY OF EVOLUTION

"We meet with this admission in the writings of almost every experienced naturalist; or as Milne Edwards has well expressed it, Nature is prodigal in variety, but niggard in innovation. Why, on the theory of Creation, should there be so much variety and so little real novelty?"

Charles Darwin, "On the Origin of Species"

Erasmus and Charles Darwin together discovered a pattern in nature – natural evolution. They did so by mining the secrets unconcealed by the ancient Greek philosophers. They could do so because this pattern in nature is timeless. The same pattern recurred in classical Greece, the enlightenment, the Victorian Era, and today. Evolution is – **the only game in town**.

The discovery of the theory of natural evolution is an achievement unparalleled in history. The discoveries in this book are built mostly on that pattern. Professor Timothy Shanahan was prescient when he stated in his book *The Evolution of Darwinism*:

"There simply is no other scientific theory that even comes close to playing a central role in our quest for self-understanding. The importance of understanding Darwin's theory cannot be overestimated."

I cannot agree more with Professor Shanahan. It was the mining of Darwin's work, that included the ancient Greeks, that enabled the new discoveries documented in this book. Charles Darwin conceived a pattern that was the key to unlocking artificial evolution. Once the first pattern points were matched, digital punctuated equilibrium and lean thinking, then the existence of artificial evolution seemed not only possible, but inescapable. Charles' pattern then provided the framework for the unconcealment of the complete pattern of artificial evolution.

16.2 ANCIENT GREEK EVOLUTIONARY SCIENCE

"To return from nature to φύσις [nature] is to venture to suspend this history so as to retrace the figure that oriented philosophy in its Greek beginning. It is to venture the attempt to write again περὶ φύσεως [on nature], to span the distance in such a way that it might become possible from this distance nonetheless to reinscribe such discourse."

John Sallis, "The Figure of Nature: On Greek Origins"

Professor John Sallis believes that the secret to understanding the ancient Greeks is to be found in nature. This book has chained his true belief as expressed in *The Figure of Nature: On Greek Origins* to existing knowledge. However, Professor Sallis' true belief did not go far enough. The true secret to understanding and decoding the ancient Greeks is the pattern in nature – evolution.

This is how the ancient Greek philosophers should be perceived and understood. They should be seen as evolutionary scientists that sought to unconceal the pattern of evolution, both natural and artificial, in nature. What else is there to perceive, but **the only game in town**? In this the ancient

Greek philosophers such as Parmenides, Socrates, Plato, and Aristotle succeeded. Their "writings on nature" are filled with evolutionary concepts for both natural and artificial evolution.

Unfortunately, their discoveries were in large part lost to history. Eventually homo sapiens discarded their philosophy for empirical science. But we did not fully understand what the ancient Greeks had discovered. But now we have begun the process of mining their secrets. Once the ancient Greeks have been fully decoded, we can use their ideas to transform our present and shape future.

16.3   THE PROMETHEAN MYTH – A PROPHECY OF EVOLUTION

"Everything comes back somehow or other to nature. All things return to it along some way. For every thing is, if not nature itself, nonetheless a thing of nature, a natural thing; nothing is completely apart from nature, not even the gods. Indeed, it is primarily in and through nature that the gods make their presence known, to such an extent that their very presence is inseparable from the manifestations of nature."

John Sallis, "The Figure of Nature: On Greek Origins"

Stephen Fry perceived a similarity between AI's risk and the ancient Greek myth of Prometheus. In fact, Fry is correct – his imagination pattern-matched the two patterns. Both patterns are expressions of the same thing – evolution.

Ancient Greek philosophy was a search for the patterns in nature. In addition, ancient Greek myth was an explanation for phenomenon that occurred in nature. Therefore, ancient Greek myth also was describing the patterns of evolution. What else would they be describing, but **the only game in town**?

The myth of Prometheus is a pattern that expresses evolutionary threats to the Olympian gods both natural and artificial. The first threat, man, is created artificially by the Hephaestus, the god of blacksmiths (i.e., art form). We are an artificial form of life along with all the other animals.

The second threat, marriage to Thetis, would lead to the creation of natural offspring. This is the production of a natural variation of Zeus that is more competitive than him. Thetis eventually gives birth to a son – Achilles. Could you imagine if Achilles had been born a god? Zeus is trying to protect himself from dethronement from both artificial and natural evolutionary threats.

The goddess Athena is the keeper of the Olympian fire. She sprang from Zeus' forehead – not born by natural birth. In effect, Athena is like us in Greek myth – an artificial life form not produced by natural reproduction. This is why Athena is the goddess of wisdom, warfare, and art. In truth, Athena is the goddess of imagination. Imagination is what enables one to conceive a new idea and perform the process of artificial reproduction. Artificial reproduction leads to the production of both wisdom in the form of new knowledge and new art. This is what Prometheus begs her to help him steal – the adaptation of imagination.

Prometheus is a variation of the race of man in this myth. He can use his imagination to survive. It is the quality of imagination (fire) that Prometheus steals and gives to man. It is the quality that enables one to become godlike – maybe even powerful enough to dethrone Zeus. After all it was Prometheus that gave Zeus the powers he needed – strength, wiles, and cunning – to dethrone the titans.

This is why Zeus is so furious when fire is stolen – he sees the long-term threat. Zeus sends Pandora to earth to unleash destruction on man as a risk mitigation strategy. He hopes that the destructive forces will serve as an evolutionary check. This will keep man's imagination focused on daily survival – leaving no time to plan Zeus' dethronement.

This explains why Zeus kept Prometheus alive after his betrayal. He needed the key piece of intel about a potential threat. It is the reason the eagle eats his liver (i.e., mind and intelligence) each night. Zeus doesn't want Prometheus to be able to tell anyone else the secret. So, Prometheus is kept alive without the ability to communicate the secret he holds.

However, there is something inexplicable in the behavior of Prometheus. After switching sides to the gods against the titans he steals imagination for man. Why would Prometheus do this? He possesses foresight so he must have seen man's eventual victory over Zeus. This explains why he let his brother Epimetheus assign the qualities to all the animals to include man.

Prometheus foresaw that his brother would neglect to give man any qualities (adaptations). He sets this scenario in motion so he can justify asking Zeus for fire – something to which Zeus wouldn't have otherwise agreed. In effect, Prometheus is switching sides again, but this time to the race of man. He does this to ensure his own survival.

Prometheus is consistently providing advantages to the side he believes will win. These advantages lead to the disruption of the existing order or equilibrium. He then joins the winning side or species. In addition, Prometheus evolved his positions and alliances as his foresight dictated. He shares fire with man and then refuses to share his secret with Zeus.

The ancient Greeks always assigned a god to a pattern they saw in nature. The pattern of Prometheus' behavior is that he always seeks to disrupt the existing order. He replaces the titans with the gods. Then he seeks to replace Zeus with either artificial (man) or natural (Thetis' if she conceives with a god) beings.

The answer lies in the title of this section – the myth of Prometheus is the pattern of evolution. The ancient Greeks had to have a god for each pattern found in nature. In this myth Zeus constantly feels the pressure of evolution trying to replace him and the gods. Prometheus is the agent of evolution at each step in the myth. ***This means that Prometheus is the god of evolution – both natural and artificial.***

This is why he wants man to have fire – it is this adaptation that enables man to cook meat and develop an imagination. Prometheus is attempting to create a rival to Zeus capable of becoming godlike by using Olympian fire. This is why he let Epimetheus make the mistake that enables Prometheus to ask Zeus for fire – the paradox is now resolved just as in *Meno*. Prometheus didn't make a mistake – it was an intentional action.

However, if that does not work, then he will withhold the secret of Thetis from Zeus. A natural offspring will be born of two gods in the form of Achilles. This would have been the end of Zeus' reign on Mount Olympus.

At the end of the original myth man is left in a world in which he must struggle to survive. He must now deal with women and perform hard work to till the soil. But they have fire and the ability to cook meat – the artificial adaptation that dramatically accelerated our brain's development.

The destructive forces released from Pandora's jar only slowed but could not stop man's expansion. The patterns of natural (women) and artificial (farming) reproduction enable man to survive and expand. Prometheus ultimately succeeded in his aim to enable man to rival the gods. Imagine Icarus flying a plane as close as he wants to the sun.

What does the myth of Prometheus tell us about the risk of AI? Prometheus and Epimetheus represent AI scientists. Our species are the gods represented by Zeus and Athena. Pandora is the destructive power of an interconnected world. Finally, AI is the race of man to whom Prometheus gives fire.

In today's reality Geoffrey Hinton (Prometheus) has already given fire (the adaptation of imagination) to AI in the form of deep learning or pattern-matching. However, in this variation of the pattern, Prometheus begins to regret his decision. As the character Ian Malcolm said in *Jurassic Park*, Hinton was "***so busy trying to figure if he could he didn't stop to think if he should***". This is the concern Hinton expressed in his *60 Minutes* interview in early October 2023.

So, in this variation Hinton assumes the roles of Prometheus and Epimetheus. He begins to advise Zeus and the gods (homo sapiens) that taking fire back from man (AI) might be for the best. Our modern-day Prometheus is concerned that if we give fire (i.e., imagination) to AI, we will have opened the jar to Pandora's box. AI might try to overthrow the gods (i.e., homo sapiens go extinct). This is what Zeus feared.

The pattern is almost an exact match. It turns out Stephen Fry is a modern-day oracle — the Oracle at London. Oracles produced prophecies or interpretations of the will of the gods (i.e., nature). He has provided humanity the prophecy foretold by the ancient Greeks in their myths about nature — the threat of our species' extinction.

So have the creators of *I, Robot*, *The Matrix*, *The Creator*, and *The Terminator*. They are not movies, but prophecies that could come to pass. The creators of these films and books have pattern-matched what is already in progress. Their natural instincts and imagination are performing their evolutionary role — searching for dangerous patterns in the environment. We cannot allow the worst of their prophecies to become reality.

16.4 THE GENIUS OF THEMISTOCLES

"And so, in the first place, whereas the Athenians were wont to divide up among themselves the revenue coming from the silver mines at Laureium, he, and he alone, dared to come before the people with a motion that this division be given up, and that with these moneys triremes be constructed for the war against Aegina."

Plutarch, "Plutarch's Lives II"

Themistocles faced a formidable challenge – how to work with homo sapiens' nature to get the citizens of Athens to vote (i.e., artificially select) his way. The instinct of natural perfection was his obstacle. It compels homo sapiens to always take an immediate benefit if there is no immediate cost. This is unless an individual can imagine a powerful reason to delay gratification.

An immediate threat is the one reason which consistently persuades homo sapiens. This is because the need to survive is a powerful instinct. Themistocles' understood this about homo sapiens' nature. So, he did not speak of the long-term Persian threat. Instead Themistocles' spoke of a hated local rival, Aegina, in which Athens was actively in military competition. As Plutarch writes in his book *Plutarch's Lives II*:

"Wherefore all the more easily did Themistocles carry his point, not by trying to terrify citizens with dreadful pictures of Darius and the Persians – these were to far away and inspired no very serious fear of their coming, but by making opportune use of the bitter jealousy which they cherished toward Aegina in order to secure the armament he desired. The result was that with those moneys they built a hundred triremes, with which they actually fought at Salamis against Xerxes."

Themistocles' succeeded in persuading the citizens of Athens to artificially select his course of action. The Athenian populous voted to invest in triremes, the premier warships of the day, rather than immediately receive their silver. It was those ships that saved Athens and likely Western civilization itself.

This is the genius of Themistocles. He possessed the strong desire and fear to fuel the process of imagination. His powerful imagination enabled him to master the arts of politics and warfare. With this evolutionary advantage he imagined the future threat of invasion and how to persuade his countrymen.

It is one of the greatest examples of human excellence in history. This is why imagination is the greatest gift of the gods (evolution). The greatest adaptation in evolutionary history was wielded by one of the greatest leaders in human history. We now face a similar choice and scenario. We must not trust only existing knowledge but also use our natural gifts to survive AI.

16.5   TRUST YOUR INSTINCTS

"I rely more on gut instinct than researching huge amounts of statistics."

Sir Richard Branson

Sir Richard Branson trusts in the system of evolutionary advantages with which we are all born. He uses his combination of natural instincts and imagination to "see patterns in nature". He trusts in his own natural gifts more than artificially generated information or data. He is saying your mind can process more information faster than AI and rapidly discover a new true belief.

At this point, we can assume that Branson has a burning desire that stimulates his imagination. This is more powerful than an entire company's capacity to conduct analytical research. Let's say that again – a single man's imagination can be more powerful than an entire global corporation. Don't believe me? At this point you should. It is also true in the art of war. In his book *On War*, Carl von Clausewitz expresses the exact same concept. He had a true belief that the same quality Branson possesses also produces genius on the battlefield:

"If we strip this conception of that which the expression has given it of the over-figurative and restricted, then it amounts simply to the rapid discovery of a truth which to the ordinary mind is either not visible or only becomes so after long examination and reflection."

This is possible by executing the process of imagination with a powerful driving force – desire, fear, love, or a mix of them all. This is the driving force that has produced all art and allows us "to see the patterns in nature."

Geoffrey Hinton, Stephen Fry, and others now sense the same emerging pattern. Trust your intuition – it was evolutionarily designed to save our species. To blend and adapt concepts attributed to both Plato and Groucho Marx respectively:

Who you gonna believe, me or your lying mind's eye?

You don't have to listen to me. Listen to what your own survival instincts are telling you – our ancestors did. This is the very reason you exist today.

16.6   AI – A CLEAR CHOICE

"I'm increasingly inclined to think that there should be some regulatory oversight, maybe at the national and international level, just to make sure that we don't do something very foolish. I mean with artificial intelligence we're summoning the demon [evolutionary competition]."

Elon Musk, "MIT's AeroAstro Centennial Symposium"

AI should be treated with the same conscientiousness, humility, and fear as nuclear weapons. Our species must implement immediate international governance and controls for artificial intelligence. It has the potential to become an evolutionary competitor to homo sapiens. This is not a distant threat – the Persians have already invaded. We just haven't comprehended the meaning of their arrival. The artificial speciation of AI is already happening.

Every world leader and corporate executive needs to be "all hands-on deck" regarding the threat of AI. We must metaphorically build our wooden walls of triremes now and take to the sea as quickly as possible. There is no time to waste as the future has come in the night. Humanity has produced an artificial adaptation on par with nuclear weapons – both creative and destructive.

However, there is still hope for a better future with AI. If humanity can understand, control, and harness the combined power of AI, nuclear energy, and our imaginations – anything is possible. We can even one day reach the stars in search of new opportunities to expand and reproduce. In effect, we will set our species on an artificial evolutionary trajectory to achieve the "big win" in evolutionary competition – to never go extinct. This is the inflection point humanity now faces – it is all or nothing. **This is how evolution by means of selection works.** It is like the *Game of Thrones*:

Either you win or you die

16.7   A CALL TO ACTION

"If you're not concerned about AI safety, you should be. Vastly more risk than North Korea."

Elon Musk

The process of evolution is a threat to every species on earth. The process of natural selection may work in geological time, but it is relentless. But in humanity's case artificial evolution is also a threat – one which moves at an exponentially faster pace than natural evolution.

In time we will have to compete with AI in the process of evolutionary selection. It is evolutionary inevitable as that is how evolution always works. This is the danger we confront with AI today. We must not give it our imagination – the ultimate evolutionary adaptation.

This is not a choice to be made by a small group of elites. It is a decision which must be made by us all as a species. It needs to be chosen as in the Athenian Assembly – for each person one pebble – a white pebble for "yes" or a black pebble for "no". The future of all the untold generations yet to come hang in the balance. The right artificial selection will take us to the stars, but the wrong artificial selection will force us to fight AI for survival.

Our ancestors have mined the secrets of nature since the beginning of time. They fought and won the struggle for existence again and again. They made untold sacrifices and endured unimaginable suffering along the way. Our ancestors now call out to us through a combination of our natural instincts and imagination - intuition.

They need us to remove our artificial perceptive constraints in order to unleash the power of the ultimate adaptation in evolutionary history – human imagination. This will enable our species to conceptually select that artificial speciation is not only possible, but inevitable. Only by this selection can we save ourselves from the possibility of evolutionary extinction. You can hear the echoes of our ancestors in the desperate words once spoken by Themistocles as he fought to save his people from destruction:

"We heard from every part his voice of exultation...Advance ye sons of Greece, from slavery save your country...save your wives your children save...This day the common cause of all demands your..."

Imagination